The Clerk

Weather Warnings for Watchers

The Clerk

Weather Warnings for Watchers

ISBN/EAN: 9783337343019

Printed in Europe, USA, Canada, Australia, Japan

Cover: Foto ©berggeist007 / pixelio.de

More available books at **www.hansebooks.com**

THE TWO OCEANS.

(1) Aërial Ocean. (2) Greatest height attained by Messrs. Glaisher and Coxwell, being 36,960 feet, or seven miles above the sea level. (3) Aërial Alps, or stratum of clouds 15,000 feet in depth. (4) Highest bird-region.

WEATHER WARNINGS

FOR

WATCHERS

BY "THE CLERK"

HIMSELF.

WITH CONCISE TABLES FOR CALCULATING HEIGHTS

"The actuating force of every wind that blows, of every mighty current that
streams through ocean depths; the motive cause of every particle of
vapour in the air, of every mist and cloud and raindrop, is
SOLAR RADIATION."—*George Warington*

THIRTEENTH THOUSAND.

LONDON

HOULSTON AND SONS

PATERNOSTER SQUARE, E.C.

1887

[The right of translation is reserved.   Entered at Stationers' Hall.]

LIST OF WORKS OF REFERENCE.

BOUTAN ET D'ALMEIDA. Cours Elémentaire de Physique.

BUCHAN, A. Introductory Text-book of Meteorology. *W. Blackwood and Sons, 1871.*

CAZIN, ACHILLE. La Chaleur. *Hachette and Co., 1868.*

CRAMPTON, REV. JOS., M.A. The Three Heavens. *W. Hunt and Co., 1876.*

CHAMBERS' Encyclopædia. *W. and R. Chambers, 1875.*

DREW, JOHN. Practical Meteorology. *Van Voorst, 1870.*

FITZROY, THE LATE ADMIRAL. Weather Book and Barometer Manual.

FLAMMARION, CAMILLE. L'Atmosphère.

GUILLEMIN, AMEDÉE. Les Forces de la Nature.

GLAISHER, J, F.R.S. Hygrometrical Tables. *Taylor and Francis, 1869.*

HARTLEY, W. N Air and its Relations to Life. *Longmans, 1875.*

HERSCHEL, SIR OHN F. W. Meteorology, from Ency. Brit. *A. and C Black, 1860.*

KAEMTZ, L. F. Complete Course of Meteorology. *Baillière, London.*

MARTIN's Natural Philosophy. *Simpkin, Marshall and Co., 1868.*

TYNDALL, JOHN, D.C.L., &c. Heat a Mode of Motion. Fifth Edition. *Longmans, 1875.*

RODWELL. Dictionary of Science. *E. Moxon and Co, 1871.*

PROCTOR. Science Byways. *Smith, Elder and Co., 1875.*

SCOTT, R. H., M.A.,F.R.S. Instructions in the Use of Meteorological Instruments, 1875.

WARINGTON, GEORGE. Phenomena of Radiation.

CONTENTS.

PREFACE TO THIRD EDITION.

THE late Admiral Fitzroy entertained the opinion that the various phenomena which go to form what we call "weather" are "measurable at any place, and that having these measurements at *various* places over a given area, such as the British Isles, we ought to be able to foresee the peculiar results as regards the direction and force of air currents which have their distinctive weather characteristics in relation to temperature, rainfall, and electrical manifestations."

A conviction of the soundness of this opinion has induced the writer to make the present compilation, in the hope that many who have hitherto avoided the subject of meteorology and the weather may find interesting matter, where before all seemed difficult and technical.

Any attempt at rigid mathematical accuracy is disclaimed at the outset; the leading principles involved in weather forecasting and storm prevision will, however, be stated in a sufficiently definite manner to divest the subject of the mystery in which it has hitherto seemed to be enshrined, and thus enable the unscientific reader to become weatherwise, and casual observers to note weather phenomena with some degree of method and precision.

On page 4 will be found a list of works which have proved useful aids in making the present compilation. The writer desires to acknowledge his indebtedness to the various authors and publishers, and especially to Mr. Strachan, for permission to quote from his able pamphlet on "Weather Forecasts, and Storm Prevision," and to reproduce the valuable table on page 37, for Calculating Heights of Mountains, from the fourth edition of his handy "Pocket Meteorological Register."

The publication of Weather Reports in the daily journals must have convinced the most indifferent that much greater importance is now attached to weather phenomena than formerly; and this conviction will be deepened when it is remembered that a Parliamentary grant of £14,000 is annually expended in support of the Meteorological Office and its seven fully organized observatories in this country, while America expends no less a sum than £80,000 annually in the pursuit of weather wisdom; and the leading nations of Europe have also established meteorological offices in suitable localities.

The balloon ascents of Messrs. Glaisher and Coxwell attracted much attention to the instruments used in estimating atmospheric phenomena, and awakened a desire to know something of the functions of a barometer, thermometer, hygrometer, &c., and especially of the classification of those important weather-warners, clouds. These subjects will be found duly noted in their order, and every phenomenon being traced to its source, Solar Radiation, it is hoped that these pages may prove generally acceptable, and be deemed not altogether unworthy of   "THE CLERK OF THE WEATHER."

WEATHER WARNINGS.

THE two great Forces of Nature are Gravitation and Heat, which always act in opposition to each other.

WEATHER is the result of the action of these forces on matter, and where one form of force is in excess of another, changes are produced which become apparent to our senses, or are indicated by suitable instruments.

THE MATTER composing the earth on which we live is of three kinds—solid, liquid, and gaseous.

THE FORCE incessantly acting on these is the radiant heat of the sun.

THE RESULTS of this incessant action are :—

1. CALORIFICATION, or Heating, which, besides being appreciable by our senses, is indicated by the THERMOMETER.

2. EVAPORATION, which alters the weight of the air indirectly, by the diffusion of aqueous vapour through it. This alteration of *weight* is indicated by the BAROMETER, the accompanying increase of moisture being indicated by the HYGROMETER.

3. RAREFACTION, which alters the *weight* of the air directly.

4. CONDENSATION, producing fog, dew, rain, hail, and snow ; all sufficiently apparent when they occur, but estimated accurately only by the Rain Gauge, or PLUVIOMETER.

5. MOTION, as in winds, which we are able to appreciate in the gentle breeze and the awful cyclone, the force and velocity of which are indicated by the ANEMOMETER.

6. ELECTRIFICATION, producing lightning, thunder, magnetic phenomena, and chemical change, respectively indicated by the ELECTROMETER, MAGNETOMETER, and OZONOMETER.

I.--CALORIFICATION.

Before considering in detail these results of the action of solar radiation on our globe, an attempt to realize the immensity of this stupendous force will materially aid in the general comprehension of the subject.

The earth is a sphere somewhat less than 8,000 miles in diameter ; and if we assume, with the gifted author * of "The Phenomena of Radiation,"—"that it is about 91,300,000 miles from the sun, and moves around it in a slightly elliptical orbit, occupying rather more than 365 days ; that its shape is globular, somewhat flattened at its two extremities ; that it rotates upon its own axis in the space of 24 hours, that axis being inclined to the annual orbit at an angle of $23\frac{1}{2}$—if we further assume that solar radiation is of such kind and quantity as it is, we are enabled to account for the total amount of light and heat the earth receives, for the superior temperature and illumination of equatorial regions, as compared with polar, with the gradations of intermediate zones, for the alternation of day and night, and the annual progression of the seasons.

"The actuating force of every wind that blows ; of every mighty current that streams through ocean depths ; the motive cause of every particle of vapour in the air of every mist and cloud and raindrop, is SOLAR RADIATION.

"The delicate tremor of the sun's surface particles, shot hither through thirty million leagues of fine intangible æther, has power to raise whole oceans from their beds, and pour them down again upon the earth. We are apt to measure solar heat merely by the sensation it produces on our skin, and think it small and weak accordingly ; a good coal fire will heat us more. But its true measure is the work it does. Judged by this standard, its immensity is overpowering. To take a single instance: the average fall of dew in England is about five inches annually ; for the evaporation of the vapour necessary to produce this trifling depth of moisture,

* George Warington, F.C.S.

there is expended *daily* an amount of heat equal to the combustion of sixty-eight tons of coal for *every square mile* of surface, or, for the whole of England, 4,000,000 tons. Compare now the size of England with that of the whole earth—only $\frac{1}{3888}$th part; extend the calculation to *rain*, as well as dew, the average fall of which on the whole earth is estimated at five feet annually, or *twelve* times greater; and then estimate the sum of $4,000,000 \times 3,388 \times 12 = 162,624,000,000$ tons, or about 3,000 times as much as is annually raised in the whole world; and we have the number of tons of coal required to produce the heat expended by the sun merely in raising vapour from the sea to give us rain during a single day.

SOLAR RADIATION.

Seeing, then, that solar radiation plays so important a part in the production of the natural phenomena classed under the head of Meteorology, a description of the mode of estimating its amount will prove interesting, and enable the reader to realize the existence of this mighty power. M. Pouillet devised for this purpose the apparatus known as the PYRHELIO-METER, which registers the power of parallel solar rays by the amount of heat imparted to a disc of a given diameter in a given time. It consists of a flat circular vessel of steel A, having its outside coated with lamp-black B. A short steel tube is attached to the side opposite to that covered with lamp-black, and the vessel is filled with mercury. A registering thermometer C, protected by a brass tube D, is then attached, and the whole is inverted and exposed to the sun, as shown at Fig. 1. The purpose of the second disc, E, is to aid in so placing the apparatus that it shall receive direct parallel rays. It is obvious that if the shadow of the upper

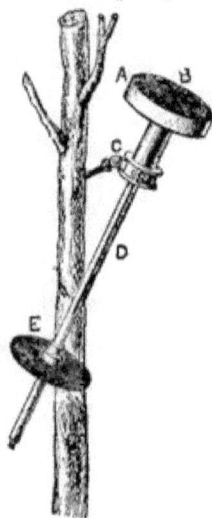

1.
Pouillet's Pyr-
heliometer. Scale
about $\frac{1}{3}$.

disc completely covers the lower one, the sun's rays must be perpendicular to its blackened surface.

"The surface on which the sun's rays here fall is known ; the quantity of mercury within the cylinder is also known ; hence we can express the effect of the sun's heat upon a given area by stating that it is competent, in five minutes, to raise so much mercury so many degrees in temperature."[*]

Sir John Herschel also designed an instrument for observing the heating power of the sun's rays in a given time, to which the title Actinometer is given.   It consists of a Thermometer with a long open scale and a large cylindrical bulb, thus combining the best conditions for extreme sensibility.   An observation is made by exposing the instrument in the shade for one minute and noting the temperature.   It is then exposed to the sun's rays for one minute, and a record of the temperature made.   It is again placed in the shade for one minute, and the mean of the two shade readings being deducted from the solar reading shows the heating power of the sun's rays for one minute of time.

The stimulus imparted to the study of this class of phenomena by the publications of Professor Tyndall's researches on Radiant Heat has induced a demand among Meteorologists for instruments capable of yielding more available indications than those just described.   This demand has been most efficiently supplied by the ingenuity of scientists and instrument makers.

2.
Herschel's
Acti-
nometer.
Scale
about ⅓.

The early form of Solar Radiation Thermometer was a self-registering maximum thermometer, with blackened bulb, having its graduated *stem*, only, enclosed in an outer tube. Errors arising from terrestrial radiation and the

* Tyndall, "Heat a Mode of Motion."

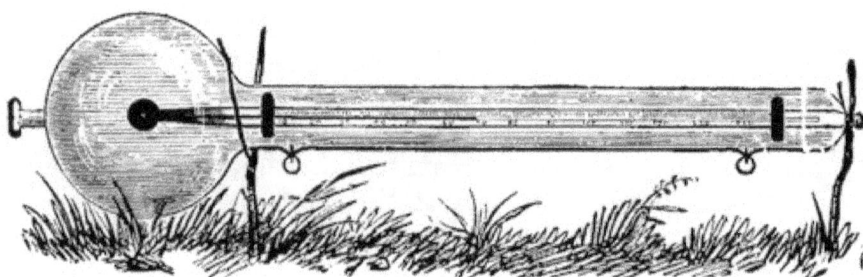

3
Improved Solar Radiation Thermometer in Vacuo.
Scale about ⅓.

variable cooling influences of aërial currents are all
obviated in the improved and patented Solar Radiation
Thermometer shown at Fig. 3, which consists of a self-
registering maximum thermometer, having its *bulb
and stem* dull-blackened, in accordance with the suggestion of the Rev. F. W. Stow, and the *whole* enclosed
in an outer chamber of glass, from which the air has
been completely exhausted. The perfection of the
vacuum in the enclosing chamber is proved by the
production of a pale white phosphorescent light, with
faint stratification and transverse bands, when tested
by the spark from a Ruhmkorff coil. Due provision is
made for this by the attachment of platinum wires to
the lower side of the tube, and when tested by a syphon
pressure gauge, the vacua have been proved to exist
to within 1-50th of an inch of pressure. It will thus
be seen that the indications are preserved from errors
arising from atmospheric currents, and from the absorption of heat by aqueous or other vapours, the
whole of the solar heat passing through the vacuum
direct to the blackened bulb. The contained mercury
expanding, carries the recording index to the highest
point, and thus is obtained a registration of the
maximum amount of solar radiation during the
twenty-four hours. The great advantage accruing
from the high degree of perfection to which this
instrument has been brought is, *uniformity* of construction, which renders the observations made at

different stations *intercomparable*. An enlarged view of
the thermometer is given at Fig. 3, showing the platinum
wire terminations, whereby the vacuum is tested. The
Rev. Fenwick W. Stow thus directs the manner in which
the solar radiation thermometer should be used :—

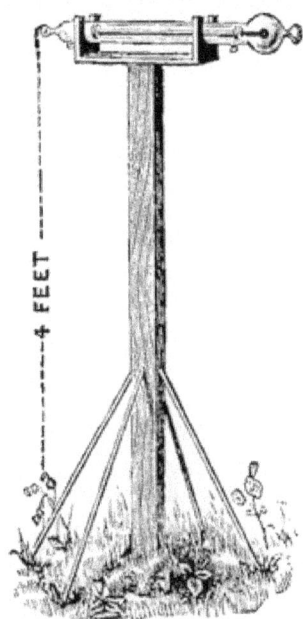

4.

Solar Radiation Thermo-
meter, black bulb and stem
in vacuo, on 4 feet stand.
Scale about $\frac{1}{20}$.

1. Place the instrument four
feet above the ground, in an
open space, Fig. 4, with its
bulb directed towards the S.E.
It is necessary that the globular
part of the external glass should
not be placed in contact with or
very near to any substance, but
that the air should circulate
round it freely. Thus placed, its
readings will be affected only
by direct sunshine and by the
temperature of the air.

2. One of the most convenient
ways of fixing the instrument
will be to allow its stem to fit
into and rest upon two wooden
collars fastened across the ends
of a narrow slip of board, which
is nailed in its centre upon a post
steadied by lateral supports
(Fig. 4).

3. The maximum temperature of the air in shade
should be taken by a thermometer placed on a stand
in an open situation. Any stand which thoroughly
screens it from the sun, and exposes it to a free circu-
lation of air, will do for the purpose.

4. The difference between the maxima in sun and
shade, thus taken, is a measure of the amount of solar
radiation.

The remarkable phenomenon recently discovered by
Mr. Crookes, in which light is apparently converted
into motion, has, at the suggestion of Mr. Strachan,
received an interesting application to meteorology. The

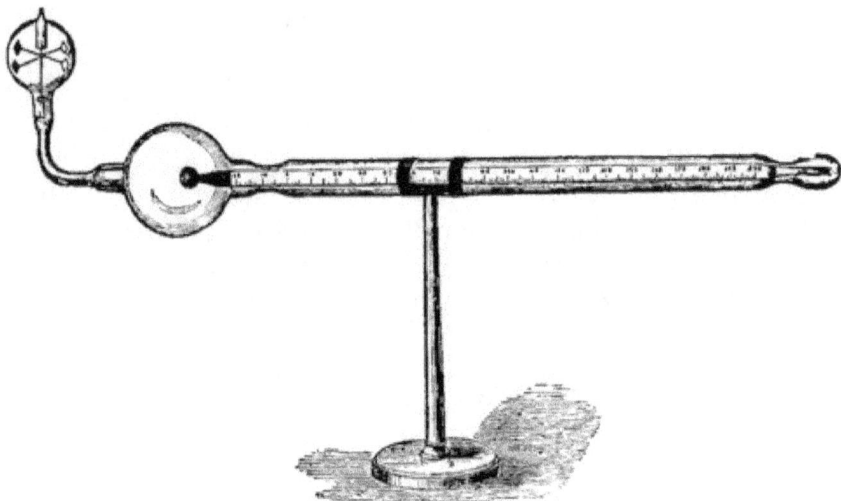

Radio-Solar Thermometer.   Scale about ¼.

arrangement is shown at Fig. 5, where a Solar Radiation Thermometer has a Crookes' Radiometer attached to it, which, in addition to forming an efficient test as to the perfection of the vacuum, will, it is hoped, aid in eventually establishing a relation between intensity of radiation, as shown by the thermometer, and the number of revolutions of the radiometer.   The instrument has so recently been devised that any positive statement as to its usefulness would be premature; it may, however, prove a valuable auxiliary to the solar thermometer, and eventually be so far improved as to become a more definite exponent of solar radiation than the thermometer.

TERRESTRIAL RADIATION.

It is an established fact, confirmed by careful experiments, that a mutual interchange of heat is constantly going on between all bodies freely exposed to view of each other, thus tending to establish a state of equilibrium.   It has further been ascertained that, as the mean temperature of the earth remains unchanged, "it necessarily follows that it emits by radiation from and through the surface of its atmosphere, on an average, the exact amount of heat it receives from the sun." This process commences *slowly* at sunset, and proceeds with great rapidity at and after midnight, attaining its maximum effect in a long night, in perfect calm, under a cloudless sky, resulting in the condensation of vapour in the form of dew, or hoar-frost, when the temperature of the surface-air is reduced to the dew-point.*

The extent to which heat thus escapes by radiation under varying conditions of sky is measured by a Self-registering Terrestrial Minimum Thermometer, the bulb of which is placed over short grass, and "a thermometer so exposed under a clear sky always marks several degrees below the temperature of the air, and its depression affords

Terrestrial Radiation Thermometer.
Scale about ⅓.

a rude measure of the facility for the escape of heat afforded under the circumstances of exposure." †

Fig. 6 shows the ordinary spherical bulb thermometer employed for this purpose, and Fig. 7 the improved Cylinder Jacket Thermometer, which, by exposing a larger surface of spirit to the air, gives an instrument possessing an amount of sensibility in no way inferior to that of mercury.

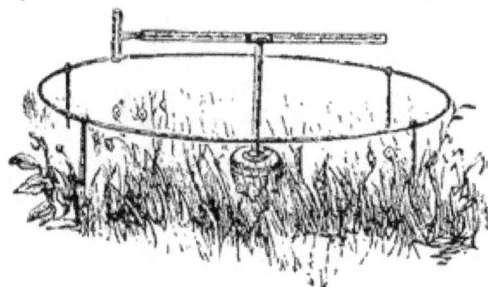

Improved Cylinder Jacket Terrestrial Minimum Thermometer. Scale about 1/12.

There is a drawback to the use of these thermo-

* See page 47.   † Herschel.

meters enclosed in outer tubes, arising from moisture getting into the outer cylinder or jacket, and frequently preventing the observer from reading the thermometer. This has recently been removed by making a perfectly ground joint of glass (analogous to a glass stopper in a bottle) as a substitute for the old form of packing at the open end of the tube, the other end being fused into contact with the outer cylinder to keep it in its place. The intrusion and condensation of moisture thus becomes impossible, while the scale is protected from corrosion or abrasion. This "ground socket" arrangement is shown at Fig. 8.

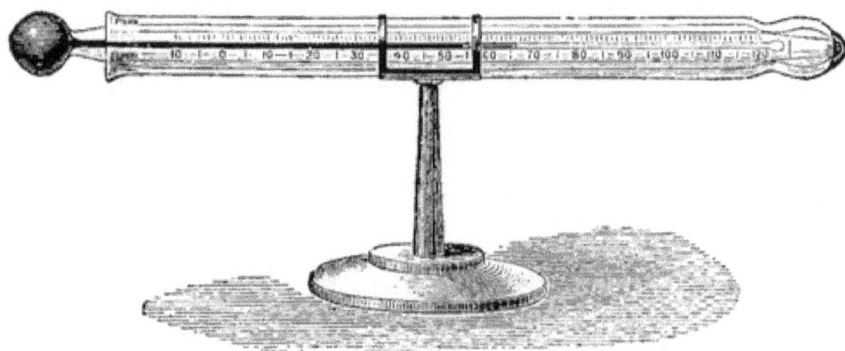

8.

Ground Socket Minimum Thermometer.   Scale about ⅓.

Radiation from the earth upwards proceeds with great rapidity under a cloudless sky, but a passing cloud, or the presence even of invisible aqueous vapour in the air, is sufficient to effect a marked retardation, as is beautifully illustrated by Sir John Leslie's Æthrioscope, shown at Fig. 9, which consists of a vertical glass tube, having a bore so fine that a little coloured liquid is supported in it by the mere force of cohesion. Each end of the tube terminates in a glass bulb containing air. A scale, having its zero in the middle, is attached to the tube, and the bulb A is enclosed in a highly polished sphere of brass. The upper bulb B is blackened, and placed in the centre of a highly-gilt and polished metallic cup, having a movable cover F. These outer metallic coverings protect the bulbs from extraneous sources of

9.
Æthrioscope.
Scale about ⅐.

heat. So long as the upper bulb is covered, the liquid in the tube stands at zero on the scale, but immediately on its removal radiation commences, the air contained in B contracts, while the elasticity of that contained in A forces the liquid up the tube to a height directly proportionate to the rapidity of the radiation.

SHADE TEMPERATURE.

Self-registering Maximum Thermometers are made in two ways. In the first, the index is a small portion of the mercurial column separated from it by a minute air bubble. The noontide heat expands the mercury

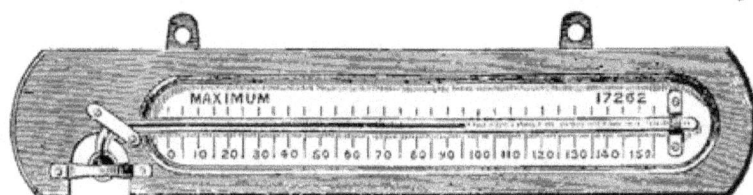

10.
Self-registering Maximum Thermometer. Scale about ⅓.

and the subsequent contraction as the temperature decreases affects only that portion of the mercury in connection with the bulb, leaving the disconnected portion to register the maximum temperature. In the second form the tube is ingeniously contracted just outside the bulb, so that the mercury extruded from the bulb by expansion cannot return by the mere force of cohesion, but remains to register the highest temperature.

There is a modification of this latter form produced by the addition of a supplementary chamber just outside the bulb and *over* the column, from which, as expansion proceeds, the mercury flows by gravitation, but into which it cannot return until, as in the other forms, the instrument is readjusted for a new observation, by unhooking the bulb end and lowering it until the mercury flows into its place.

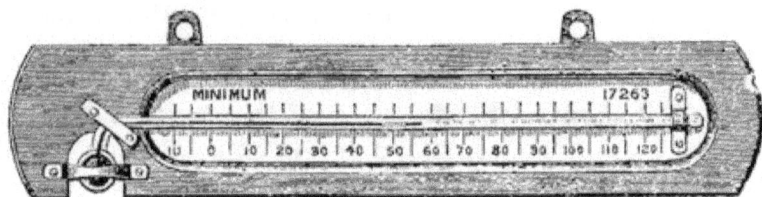

11.

Self-registering Minimum Thermometer.   Scale about ⅓.

Self-registering Minimum Thermometers are of two kinds, — spirit and mercurial.   Fig. 12 shows one of Rutherford's Alcohol Minimum Thermometers, which will be seen to

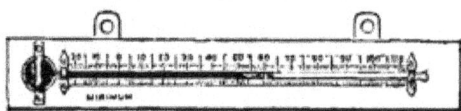

12.

Self-registering Minimum Thermometer.
Scale about ⅓.

consist of a bulb and tube attached to a scale, which latter may be either of wood, glass, or metal.   The tube contains an index of black glass.

The Thermometer is "set" for observation by slightly raising the bulb end until the index slides to the extreme end of the column of spirit.   It is then suspended in the shade with the bulb end a little lower than the other.   The contraction of the spirit consequent on a fall of temperature draws the index back, but a subsequent expansion does not carry it forward, it remains at the lowest point to which the spirit has contracted to register the minimum temperature.   A very useful modification of this instrument is made for gardeners and general horticultural purposes, in which the scale is of cast zinc with raised figures, which being filed off flush after the whole has been painted of a dark colour are easily legible at a little distance.

The advantage of alcohol for the indication of *very* low temperatures is that it has never been frozen.*

Fig. 13 shows a set of Maximum and Minimum and Wet and Dry Bulb Thermometers, with incorrodible

---

* Mercury freezes at - 39° F.

B

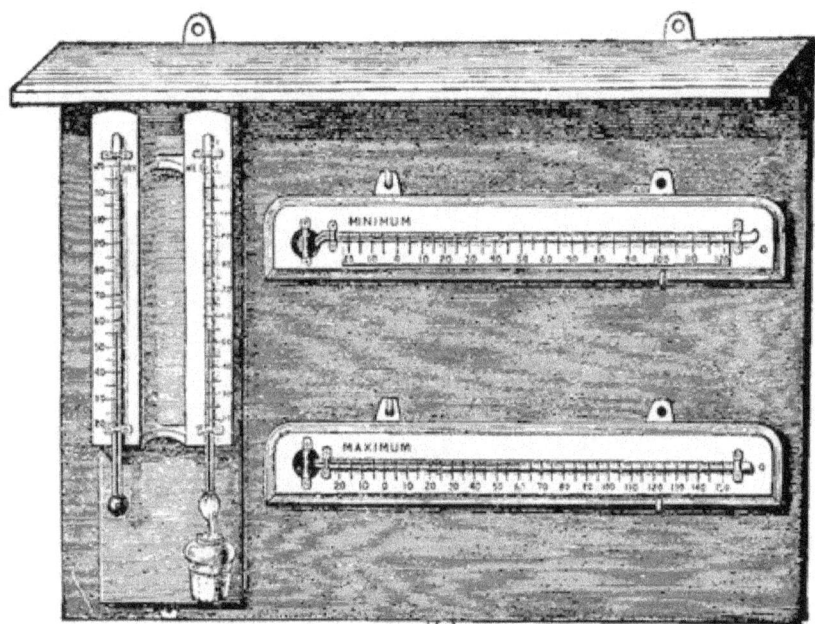

Standard Set of Instruments on Screen.   Scale about ¼.

porcelain scales, suspended on a mahogany screen.
Instruments of this quality are generally engine-divided
on the stem, and if, in addition to this, they are verified
by comparison with standard instruments at the Kew
Observatory, they may be regarded as standards, and
employed for accurate scientific observations.

Six's Self-registering Thermometer consists of a long
tubular bulb, united to a smaller tube more than
twice its length, and bent twice, like a syphon, so that
the larger tube is in the centre, while the smaller one
terminates at the top, on the right hand, in a pear-shaped
bulb, as shown in the cut (Fig. 14).   This bulb, and the
tube in connection with it, are partly filled with spirit;
the long central bulb and its connecting tube are completely filled, while the lower portion of the syphon is
filled with mercury.   A steel index, prevented from
falling by a hair tied round it, to act as a spring, moves
in the spirit in each of the side tubes.   The scale on
the left hand has the zero at the top, and that on the right

at the bottom. When setting the instrument, the indices are brought into contact with the mercury by passing a small magnet down the outside of each tube. Then, should a rise of temperature take place, the spirit in the central bulb expands, forcing down the mercury in the left hand tube and causing it to rise in the right, and *vice versâ* for a diminution of temperature.

It should be always used and carried upright, and the indices should be drawn *gently* down by the magnet into contact with the mercury; and, when a reading is taken, the ends of the indices nearest the mercury indicate the maximum and minimum temperatures which have been attained during the stated hours of observation.

Six's form of thermometer has been extensively used for ascertaining deep sea temperatures.

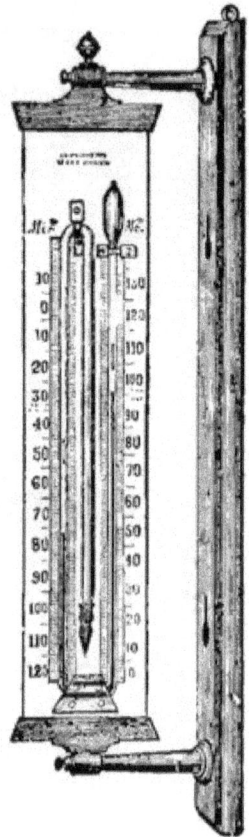

14.
Six's Thermometer.
Scale about ⅓.

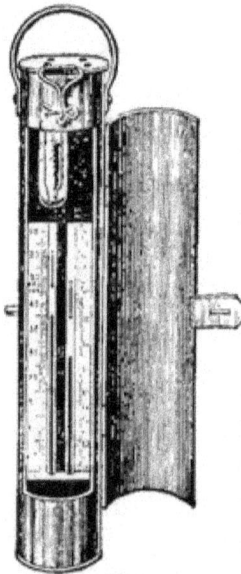

15.
Deep Sea Maximum and Minimum Registering Thermometer.
Scale about ¼.

Evaporation and the mechanical action of winds keep up a constant circulating motion of the ocean, the currents of which tend to equalize temperature. The most important of these is known as the Gulf Stream, taking its name from the Gulf of Mexico, out of which it flows at a velocity sometimes of five miles an hour, and in a width of not less than fifty miles. It has an important effect on the climate of Great Britain, and of all lands subject to its influence, its temperature as it leaves the Gulf of Mexico being 85° F., diminishing to 75° off the coast of Labrador, and still further as it nears northern latitudes. Observations on the temperature of the ocean are therefore included in the scope of meteorology, and are ascertained by the use of thermometers of special construction (Fig. 15). In the earlier experiments made for ascertaining the temperature of the ocean at a depth of 15,000 feet, where the pressure is equal to three tons on the square inch, it was found that a considerable error occurred in the indications in consequence of this enormous pressure; accordingly the central elongated bulb of the ordinary Six's Thermometer (see page 19) is shortened and enclosed in an outer bulb nearly filled with spirit, which, while effectually relieving the thermometer bulb from undue pressure, allows any change to be at once transmitted to it, and thus secures the registration of the exact temperature. The arrangement possesses the further advantage of making the instrument stronger, more compact, and more capable of resisting such comparatively rough treatment as it would receive on board ship.

The honour of constructing the first thermometer, which was an Air and Spirit Thermometer, is ascribed to Galileo; it assumed a practical shape in 1620, at the hands of Drebel, a Dutch physician. Halley substituted mercury for spirit in 1697; Réaumur improved the instrument in 1730, and Fahrenheit in 1749. More recently the instrument has been perfected by the scales being graduated on the actual stem of the instrument. For

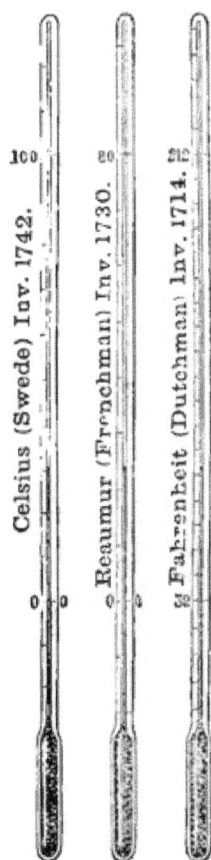

16
Comparison of Thermometer Scales.
Scale about ⅓.

many years it was exclusively used by chemists and men of science; it afterwards received numerous applications in the arts and manufactures; and is now considered an essential in every household.

Thermometers are instruments for measuring temperature by the contraction or expansion of fluids in enclosed tubes. The tubes, which are of glass, have spherical, cylindrical, or spiral bulbs blown on to one end; they have also an exceedingly fine bore, and when mercury or spirit is enclosed in them these fluids, in contracting and expanding with variations of temperature, indicate degrees of heat in relation to two fixed points—viz., the freezing and boiling points of water. Care is taken to exclude all air before sealing, so that the upper portion of the tube inside shall be a perfect vacuum, and thus offer no resistance to the free expansion of the mercury. In graduating, or dividing the scales, the points at which the mercury remains stationary in melting ice and boiling water are first marked on the stem, and the intervening space divided into as many equal parts as are necessary to constitute the scales of Fahrenheit, Réaumur, or Celsius, the last being known as the Centigrade (*hundred steps*) scale, from the circumstance of the space between the freezing and boiling points of water being divided into one hundred equal parts (Fig. 16).

GRADUATION OF THERMOMETERS.—When the fluid (either mercury or spirit) has been enclosed in the hermetically sealed tube, it becomes necessary, in order that its indications may be comparable with those of other instruments, that a scale having at least two fixed points should be attached to it. As it has been found that the temperature of melting ice or freezing water is always constant, the height at which the fluid *rests*

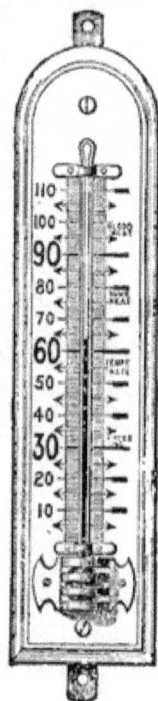

17.
"Legible"
Scale Thermometer.
Scale about ⅓.

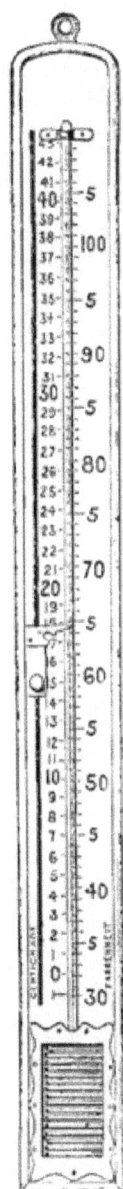

18.

Gridiron-bulb
Thermometer.
Scale about ¾.

in a mixture of ice and water has been chosen as one point from which to graduate the scale. It has been also found that with the barometer at 29·905 the boiling-point of water is also constant, and when a thermometer is immersed in pure distilled water heated to ebullition, the point at which the mercury remains immovable is, like the freezing-point, carefully marked; the tube is then calibrated and divided as shown in Fig. 16.

The zero of the scales of Réaumur and Centigrade is the freezing-point of water, marked, in each case, 0°, while the intervening space, up to the boiling-point of water, is divided, in the former case, into 80 parts, and in the latter to 100°.

In the Fahrenheit scale, the freezing-point is represented at 32°, and the boiling-point at 212°, the intervening space being divided into 180°, which admits of extension above and below the points named, a good thermometer being available for temperature up to 620° Fahr.

The use of the Réaumur scale is confined almost exclusively to Russia and the north of Germany, while the Centigrade scale is used throughout the rest of Europe. The Fahrenheit scale is confined to England and her colonies, and to the United States of America.

Circumstances sometimes arise in which it becomes necessary to convert readings from one scale into those of the others, according to the following rules:—

1. To convert Centigrade degrees into degrees of Fahrenheit, multiply by 9, divide the product by 5, and add 32.

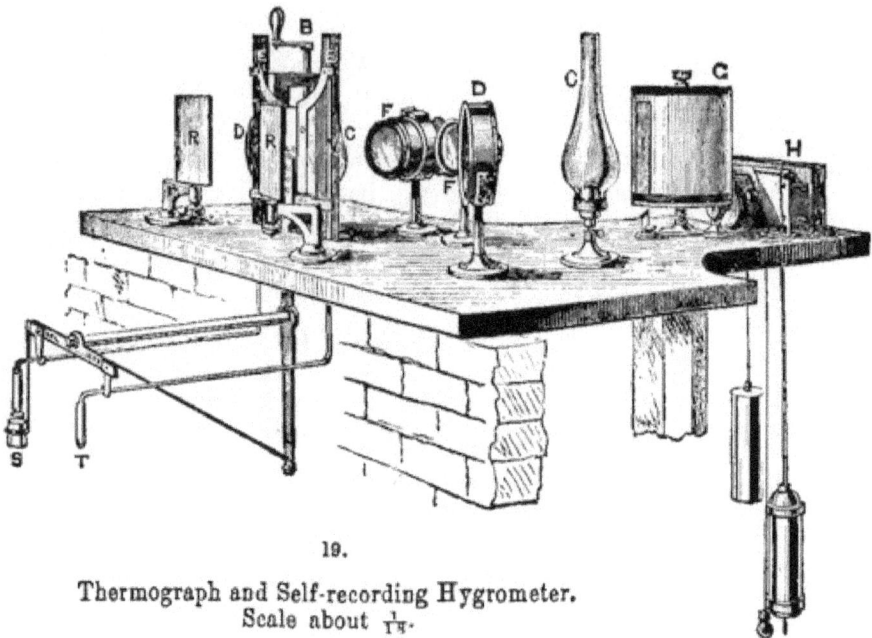

19.

Thermograph and Self-recording Hygrometer.
Scale about $\frac{1}{14}$.

2. To convert Fahrenheit degrees into degrees of Centigrade, subtract 32, multiply by 5, and divide by 9.

3. To convert Réaumur degrees into degrees of Fahrenheit, multiply by 9, divide by 4, and add 32.

4. To convert Réaumur degrees into degrees of Centigrade, multiply by 5 and divide by 4.

For the production of *continuous* records, the Meteorological Committee of the Royal Society have adopted an instrument called a Thermograph, or self-recording wet and dry bulb thermometer, which is largely aided by photography. The bulbs of the thermometers are necessarily placed in the open air, and at a suitable distance from any wall or other radiating surface; the tubes are of sufficient length to admit of their being brought inside the building, in due proximity to the recording apparatus placed in a chamber from which daylight is rigidly excluded.

The essential conditions in such an apparatus are:—
1. A means of denoting the height of the mercurial column in the stem of a thermometer in relation to a fixed horizontal line. 2. A time scale denoting the exact moment at which the atmosphere reached the temperature indicated by the mark. 3. As the marks are produced chemically, and not mechanically (as in the Anemograph), a *dark* room.

A description of the drawing on page 23 will best show how very efficiently, through the ingenuity of Mr. BECKLEY, these conditions have been obtained :—S, wet bulb thermometer; T, atmospheric thermometer; B, screw for adjusting thermometers; C C, paraffin lamps or gaslights; D D, condensers, concentrating the light on the mirrors R R; R R, mirrors reflecting light through air-speck in thermometers V V; E E, slits through which light passes from mirrors R R; F F, photographic lenses, producing image of air-speck from both thermometers on cylinder G; G, revolving cylinder or drum carrying photographic paper; H, clock, turning cylinder G round once in 48 hours; I, shutter to intercept light four minutes every two hours; leaving white time-line on developing latent image.

II.—EVAPORATION.

Solar heat rarefies the air by driving its particles asunder; it also vaporises water from the surface of river, lake, and ocean, diffusing the vapour through the atmosphere.

Great interest attaches to the subject of Evaporation, on account of its connection with rainfall and water supply. It is to be regretted, therefore, that the results hitherto obtained in the endeavour to measure its rate and quantity do not merit much confidence as regards their applicability to the evaporation occurring in nature, owing to the exceptional manner in which the observations have been made.

There is this uncertainty about evaporation, that all

the experiments relate to that taking place from an exposed water surface of a, comparatively speaking, infinitesimally small area, and can therefore have but a very partial applicability to the conditions occurring in nature. There are two main reasons for this statement. Firstly, the proportion of the surface of the land on the earth which is covered with lakes and rivers is very limited, and the experiments above indicated throw no light on the evaporation from the soil. Secondly, the evaporation from the surface of a small atmometer erected on the ground, with comparatively dry air all around it, is certainly very different from that which would take place from an equal area in the centre of a large water surface, such as a lake.

It is of course easy to make experiments on the evaporation from the soil by means of a balance atmometer, but in order that these should possess a practical value, the investigation must be extended so as to include a wide variety of soils, &c., &c. As regards the second point which has been raised, it is recommended by the Vienna Congress to erect atmometers in the centre of water surfaces; but it is not a very easy matter to conduct such experiments with accuracy, owing to the risk of in-splashing from waves.

BABINGTON'S ATMIDOMETER measures evaporation from water, ice, or snow, and in form resembles a hydrometer, with the difference that the stem bears a scale graduated to

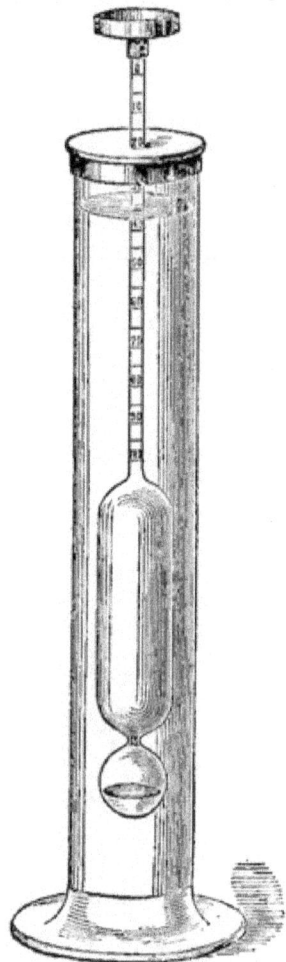

20
Atmidometer.
Scale about ½.

grains and half grains, and is surmounted by a light, shallow copper pan. When in use, the hydrometer-like instrument is immersed in a glass vessel having a hole in the cover, through which the stem protrudes. The copper pan is then placed on the top, and sufficient water, ice, or snow placed therein to sink the stem to the zero of the scale. As the evaporation proceeds, the stem rises; and, if the *time* of commencing the experiment is noted, the rate as well as the amount of evaporation is indicated in grains.

III.—RAREFACTION.

The diffusion of aqueous vapour through the air and the rarefying influence of heat jointly effect an alteration in the *weight* of the atmosphere. This alteration of weight is determined by the Barometer, an instrument invented by TORRICELLI, in 1643, and in so perfect a form that in its essential features it has not been superseded.

The mode of construction is illustrated by Figs. 21 and 22. It consists in hermetically sealing a glass tube about three feet long and filling it with mercury. The finger is placed over the open end of the tube, which is then inverted and placed in a cistern of mercury and the finger withdrawn. The left-hand figure shows the result; the mercury is seen to fall some three or four

21.
Construction of Barometer. Scale about $\frac{1}{18}$.

22.
Construction of Barometer. Scale about $\frac{1}{18}$.

inches, leaving an empty space at the top of the tube, which is called the "Torricellian vacuum."

The mercury is prevented from falling lower than is shown, by the external pressure of the atmosphere on the cistern. The *weight* of this column, therefore, represents the *weight* or pressure of a corresponding column of air many miles in height; and so close is the relation between the column of mercury and the external air that the *height* of the former changes with the slightest variation in the *weight* of the latter, and the instrument thus becomes a measure of the weight of the air, from which property its name is derived, the Greek words *baros* and *metron* signifying respectively "weight" and "measure."

When the mercury in the barometer tube falls, that in the cistern rises in corresponding proportion, and *vice versâ*, so that there is an ever-varying relation between the *level* of the mercury in the tube and the mercury in the cistern, which affects the accuracy of the readings. In M. Fortin's cistern this difficulty is obviated by the use of a glass, with flexible leather bottom and a brass adjusting screw, as shown in the cut. Through the top of the cistern is inserted a small ivory point, the lower end of which corresponds with the zero of the scale ; and, to secure uniformity, the level of the mercury in the cistern should be adjusted by the screw at each observation, until the ivory point *appears* to touch its own reflection on the surface. The reading is then taken.

23.

Fortin's
Cistern.
Scale about
⅛.

In making barometric observations for comparison with others, it is necessary that all should be reduced to the common temperature of 32° F., and for this purpose tables have been calculated which will be found to save much time.

Tables also for reducing observations of the barometer to sea-level, an operation equally indispensable with the

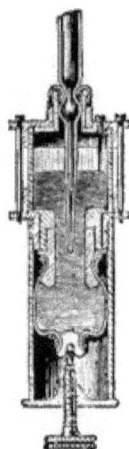

24.
Error of
Capillarity.
Scale about ½.

other corrections to make the readings intercomparable, have been published by direction of the Meteorological Committee.

For the British Isles the mean sea-level at Liverpool has been selected by the Ordnance Survey as their datum, and the height of any station may be ascertained by first noting the nearest Ordnance Bench Mark thus ↑ , and purchasing that portion of the Ordnance map which includes the station, near to which the Bench Mark will be found with the height above sea-level duly entered. The levellings made for railways will also furnish the desired information. Failing both these, the observer should select two or more of the stations nearest his locality for which official Meteorological Reports are published daily in the *Times* and other journals ; and taking observations of his barometer at 8 a.m., for a few weeks, should compare them with the mean of the observations at those stations. The comparison should be omitted when the barometer pressure is not steady.

A Standard Barometer is constructed on FORTIN's principle, and should have its tube about half an inch bore, enclosed in a brass body having at its upper end two vertical openings, in which the vernier works. The mercury is seen through these openings, aided by light reflected from a white opaque glass reflector let into the mahogany board

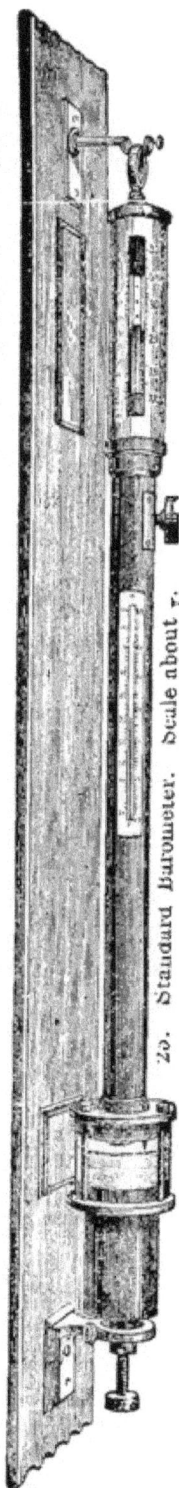

25. Standard Barometer. Scale about ⅐.

behind. The scale is divided on one side into English inches and 20ths, and may have on the other French millimetres, the vernier enabling a reading to be taken, in each case respectively, of 1-500th of an inch and 1-10th of a millimetre. In making the instrument, the mercury is boiled in the tube, to ensure the complete exclusion of air and moisture; while FORTIN'S principle of cistern ensures a constant level from whence to take the readings. A sensitive thermometer with scale, engine-divided on stem, is attached to the brass mount, which is perforated to admit the attenuated bulb of the thermometer into absolute contact with the glass tube of the barometer, to ensure its indicating the same temperature as the contained mercury. The instrument is suspended by a ring from a brass bracket attached to a mahogany board, and the lower end passes through a larger ring having three screws for adjusting it vertically.

A "reading" is taken in the following manner:—1. Note the temperature by the attached thermometer. 2. Raise or lower the mercury in the cistern by turning the screw underneath until the reflected image of the ivory point on the mercury *seems* to be in contact with the ivory itself. By the milled head at the side, the vernier is adjusted until its lower edge just touches the top of the mercurial column; the scale and vernier then indicate the height of the barometer in inches, 10ths, 100ths, and 1000ths.

High-class instruments, such as that here described, yield *exact* readings; but, in order to note them accurately, it is important that the eye, the zero edge of the vernier, the top of the mercurial column, and the back of the vernier should be in the same horizontal plane; conditions which may be obtained after some practice.

The accompanying illustration shows a form of barometer which, though not much used in this country, is deservedly popular on the Continent as a standard station barometer. It is called a Syphon Barometer, and was designed by Gay-Lussac. The

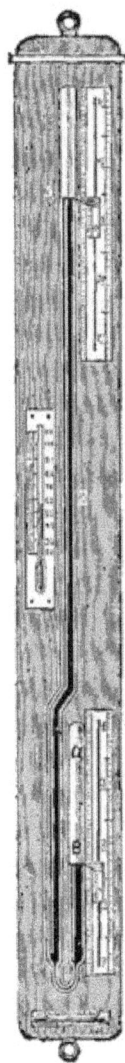

26.

Syphon
Barometer.
Scale
about $\frac{1}{12}$.

open end of the tube is bent up in the form of
a syphon, the short limb being from six to
eight inches long; it is furnished with metal
scales and verniers, and is mounted on a
mahogany board with attached thermometer.

These barometers require no correction for
capillarity or capacity, each surface of
mercury being equally depressed by capillary
attraction, and the quantity of mercury
falling from the long limb occupies the same
space in the short limb. The usual correction for temperature must, however, be
applied. A scale of inches, measured from a
zero point taken near the bend of the tube,
furnishes the means of measuring the long
and short columns. The difference of
readings is the height of the barometer.

The VERNIER is a movable scale for sub-
dividing parts of a fixed scale, and was first
applied to that purpose by its inventor, M.
PIERRE VERNIER, in 1630. In the barometer
the parts to be divided are inches, which by
the aid of this invention are subdivided into
10ths, 100ths, and 1000ths.

Fig. 27 shows the scale of a standard
barometer divided into ½-10ths, or ·05 of an
inch. The Vernier C D is made equal to 24
of such divisions, and is divided into 25 equal
parts, from whence it follows that one
division on the scale is 1-25th of ·05 larger
than one on the vernier, so that it shows
a difference of ·002 of an inch. The vernier
reads from ·0, or zero, upwards; D, therefore, indicates the top of the mercurial
column.

In Fig. 27, zero on the vernier is exactly
in line with 29 inches and 5-10ths of the fixed scale;
the reading, therefore, is 29·500 inches. The vernier

line *a* falls short of a division of the scale by ·002-inch; *b*, by ·004; *c*, by ·006; *d*, by ·008; and the succeeding line by ·010. If the vernier be adjusted to make *a* coincide with *z* on the scale, it will have moved through ·002 - inch; and if 1 on the vernier be moved to coincide with *y* on the scale, the space measured will be ·010-inch. Consequently, the figures 1, 2, 3, 4, 5, on the vernier, measure 100ths, and the intermediate lines even 1000ths of an inch. In Fig. 28 the zero of the vernier is between 29·65 and 29·70 on the scale.

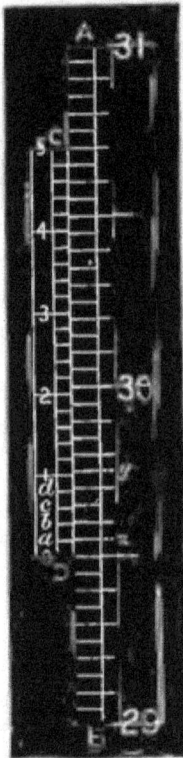

27.
The Vernier.

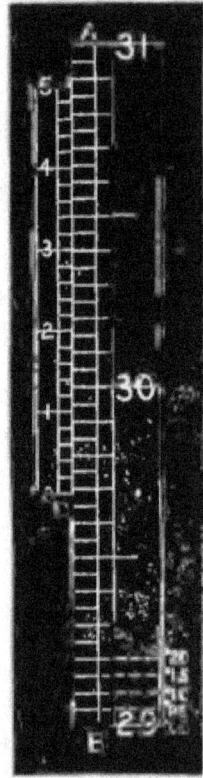

28.
The Vernier.

Glancing up the vernier and scale, the second line above 3 will be found in a direct line with one on the scale; this gives ·03 and ·004 to add to 29·65, so that the actual reading is 29·684. In those instances where no line on the vernier is found *precisely* to coincide with a line on the scale, and doubt arises as to which to select from two equally coincident lines, the rule is to take the intermediate 1000th of an inch.

For household and marine barometers such minute subdivisions of the scale are unnecessary, and the scales of such instruments are therefore divided only to 10ths, and the verniers made only to read to 100ths of an inch, which is effected by making the vernier 9-10ths or 11-10ths of an inch long, and dividing it into 10 equal parts.

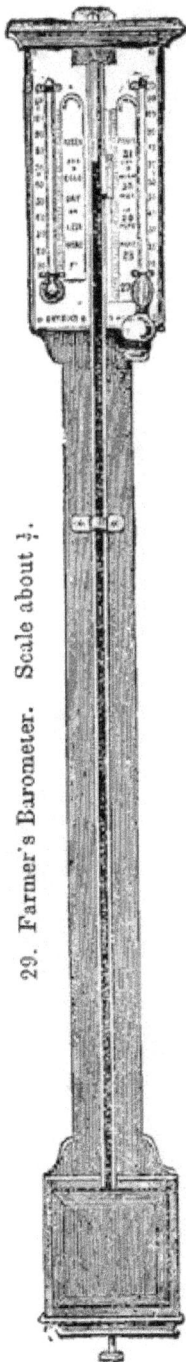

29. Farmer's Barometer. Scale about ⅓.

In "taking a reading" it is important that it should be done as quickly as possible, as the heat from the body and the hand is sufficient to interfere with that accuracy which is necessary where the intention is to compare the readings with those made by other observers. This facility is soon acquired by a little practice.

Pediment Household Barometers, though not so imposing in appearance as the Wheel Barometer, yield direct readings without the intervention of the mechanical appliances necessary for moving a needle over an extended dial. Their mountings are for the most part in oak, walnut, and other woods, the scales are of ivory, porcelain, or enamelled glass, and in their graduation due regard is paid to the relative proportions of cistern and tube, so that the conditions essential to the production of a Standard Barometer are very closely attained. In common with other barometers, it should hang in the shade in a vertical position, so that light may be seen through the tube. As a purchaser would receive it in what is called a "portable" state, it will be necessary on first suspending it to take the pinion key, fit it on the square-headed pin at the bottom of the instrument, and turn gently to the left till the screw stops. The effect of this is to lower the base of the cistern, and allow the mercury in the tube to fall to its proper level. The key should then be replaced for use in moving the vernier. To make this kind of Barometer portable for

travelling it should be un-
hung, *very* gradually sloped
until the mercury is at the
top of the tube, when, the
instrument being upside
down, the base of the cistern
is screwed up by turning
the pinion key gently to the
right until it stops. Care
should be taken to avoid
concussion, and to have the
cistern end always upper-
most, or the instrument
lying flat.

Fig. 29 shows a useful form
of barometer for the farmer,
combining as it does three
instruments in one, for the
thermometer on the right
hand of the scale having
its bulb covered with muslin
kept moist by communica-
tion with a cistern of water
enables the two thermo-
meters to be employed as a
Hygrometer, the use of
which is described at page
50. This barometer should
be suspended in a place
where it will be exposed as
much as possible to the external air, but not in sunshine.

In Wheel Barometers the
varying height of a column
of mercury is shown by the
movement of a needle on a
divided circular dial, by
adopting the syphon form of

C

Wheel Barometer, Scale
about ⅛.

barometer tube, concealed behind the dial and frame.
An iron or glass float sustained by the mercury in the
open branch (Fig. 31) is suspended by a counterbalance
a *little* lighter than itself. The axis of the pulley has
the needle attached to it, and consequently moves the
needle with the rise and fall of the mercury. It is obvious,
therefore, that if the atmospheric pressure increases the
float falls and the needle turns to the right, and if it
diminishes the needle turns in the opposite direction.
The divisions on the scale represent inches, tenths, and
hundredths in the rise and fall of a column of mercury,
and these can be read with great
facility, as one inch occupies the
space of six or more on this very
open scale, according to size of
dial (Fig. 30). The wording is
arbitrary, and indicates the *probable* weather that may be expected.

Important improvements have
recently been effected in this form
of household barometers, so that
they may be recommended as
good weather indicators where facility of reading is a desideratum.

Since the more scientific " Pediment" has attained so high a
degree of popularity, a certain
amount of unmerited obloquy has
attached itself to the Dial or Wheel
Barometer invented by Dr. Hook.
It must be conceded that the standard form of pediment barometers
in which the height of the mercury
is seen at a glance is more strictly
an "instrument of precision," but
it should not be forgotten, although
a delicate mechanism intervenes
between the mercury and the

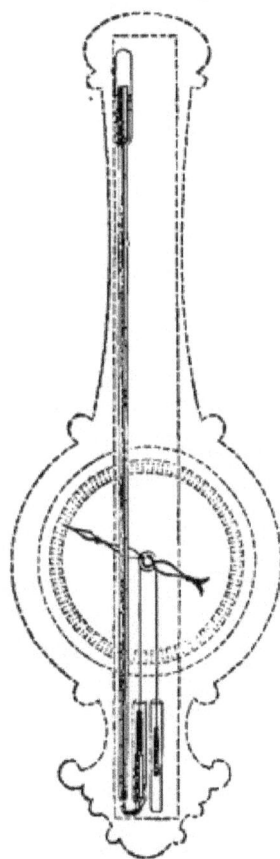

31.
Mechanism of Wheel
Barometer.
Scale about ⅓.

observer, it is so arranged that a tenth of an inch rise or fall causes a movement of the index over an inch of space.

The Aneroid Barometer indicates variations in atmospheric pressure by the elevation and depression of the sides of an elastic metallic box from which the air is exhausted and which is kept from complete collapse by a powerful spring. In cases where *extreme* accuracy is not indispensable, the portability and sensibility of this instrument recommend it for use by tourists and fishermen. It is " quick in showing the variations of atmospheric pressure."* " The Aneroid readings may be safely depended upon."† " Its movements are always consistent."‡ " Atmospheric changes are indicated *first* by the Aneroid."§ It is

32.
Aneroid Barometer. Full size.

especially adapted for determining mountain altitudes, some being furnished with a scale of feet, enabling the observer to read off the height by direct observation, and if adjusted once a year by comparison with a mercurial standard is quite trustworthy. It is fully described in a small pamphlet entitled " The Aneroid Barometer : How to Buy, and How to Use it," by a Fellow of the Meteorological Society.

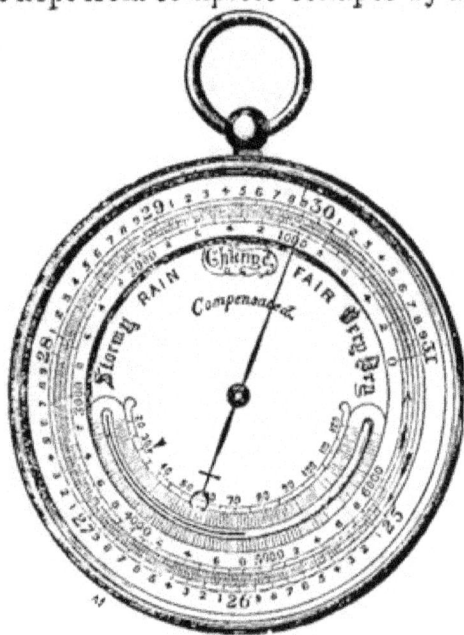

* Admiral Fitzroy.
† James Glaisher, Esq., F.R.S.
‡ James Belville, Esq., Royal Observatory, Greenwich.
§ Sir Leopold McClintock.

By a suitable arrangement of clockwork, revolving a cylinder bearing prepared paper, the aneroid barometer forms an admirable self-recording instrument, showing at a glance the height of the barometer: whether it is falling or rising, for how long it has been doing so, and at what rate the change is taking place, whether at the rate of 1-10th per hour, or 1-10th in twenty-four hours—facts which can only be obtained by very frequent and regular observations from an ordinary barometer, but which are nevertheless essential to a reliable " weather forecast."*

The height of mountains may also be determined by the temperature at which water boils, as this depends on the pressure of the atmosphere, and according to Deschanel, "just as we can determine the boiling - point of water when the external pressure is given, so if the boiling-point be known we can determine the external pressure," and as this varies with the elevation above sea-level, the boiling-point of water also varies.

These facts induced Wollaston to attempt the determination of heights of mountains by an apparatus which he called the Barometric Thermometer, subsequently modified by Regnault and called a Hypsometer, but now more generally known as a Boiling-point Thermometer.

A portable form of boiling-point thermometer is shown at Fig. 33, which is much used by Alpine travellers, and forms a trustworthy check on the aneroid and barometer.

33.
Boiling-point
Thermometer.
Scale about ⅓.

* The Aneroid Barometer: How to Buy and How to Use it. By a Fellow of the Meteorological Society. Post free for six stamps, from any bookseller or optician.

CONCISE TABLES FOR CALCULATING HEIGHTS BY MEANS OF BAROMETER OR ANEROID, AND ALSO BY THE BOILING-POINT THERMOMETER.

Boiling-point of Water for pressure in next col.	Barr. at lower Station. In.	BAROMETER AT UPPER STATION.—INCHES.															
		30	29	28	27	26	25	24	23	22	21	20	19	18	17	16	15
213·78	31	859	873	889	905	921	939	957	977	998	1020	1043	1068	1095	1124	1155	1188
212·13	30	......	888	904	920	937	955	974	994	1015	1038	1062	1087	1115	1144	1176	1210
210·43	29	......	......	919	936	953	971	991	1012	1033	1056	1081	1107	1135	1165	1198	1233
208·67	28	......	......	......	952	970	989	1009	1028	1051	1075	1100	1127	1156	1187	1220	1257
205·87	27	Factor A.				988	1007	1028	1050	1073	1097	1122	1150	1180	1211	1246	1283
205·01	26						1027	1048	1070	1093	1118	1145	1173	1203	1236	1271	1309
203·09	25	Height	D.					1069	1092	1116	1141	1169	1198	1229	1262	1299	1338
201·11	24	in 1,000	addi-	Lat.	C.				1115	1140	1166	1194	1224	1256	1290	1327	1367
199·05	23	feet.	tive.							1164	1191	1220	1251	1284	1319	1358	1399
196·92	22					M n.	Factr.			......	1218	1248	1280	1314	1350	1390	1433
194·71	21	2	5	0°	+2·7	Tem.	B.					1278	1310	1346	1383	1424	1469
192·41	20	4	11	10	+2·5	10°	0·951						1343	1380	1419	1461	1507
190·00	19	6	17	20	+2·0	20	0·973							1416	1457	1500	1548
187·50	18	8	23	30	+1·4	30	0·996					Factor A			1497	1542	1592
184·87	17	10	30	40	+0·6	40	1·018									1588	1639
182·10	16	12	37	45	0·0	50	1·040										1690
179 20	15	14	44	50	−0·5	60	1·052										
		16	52	60	−1·3	70	1·084										
		18	60	70	−2·0	80	1·127										

RULE I.—If the temperature of boiling water be observed at either or both Stations, find the equivalent pressure in the 2nd column, and calculate the height as for barometer.

RULE II.—The readings of the Barometer being corrected and reduced to 32° F., multiply the difference of pressure between the Stations by factor A, found in line with pressure at lower Station, and under that at upper Station; multiply again by factor B, corresponding to the mean temperature of the air at the Station; apply as many times C as there are thousand feet in the height, corresponding to the latitude; and add D, the correction for gravity.

EXAMPLE.—At the top of Snowdon, lat. 53° N., an aneroid read 26·48, correction —0·18, the pressure at sea-level was 29·91 the temperature of the intermediate air was 57°; find the height.

Lower Station   29·91 inches.
Upper   ,,   26·30

                  3·61
Factor A   933

                 1083
                 1083
                 3249

                 3368   (neglecting decimals.)
Factor B   1·055

                16840   N.B.—In taking out the quantities, if accuracy is aimed at, it
                16840   will be necessary to proportion for parts in the usual manner
                 3368   with such Tables.

                 3553
Cor. C = 3 × 1 = − 3
Cor. D   + 10

Height   3560 feet.

The illustration (Fig. 33) shows the instrument with the telescopic tube drawn out for use, and the thermometer surrounded by the vapour of boiling water. The lamp is protected from wind by a perforated japanned tin case covered with wire gauze. When the boiler is charged and the lamp ignited the mercury ascends, and the point at which it becomes stationary shows

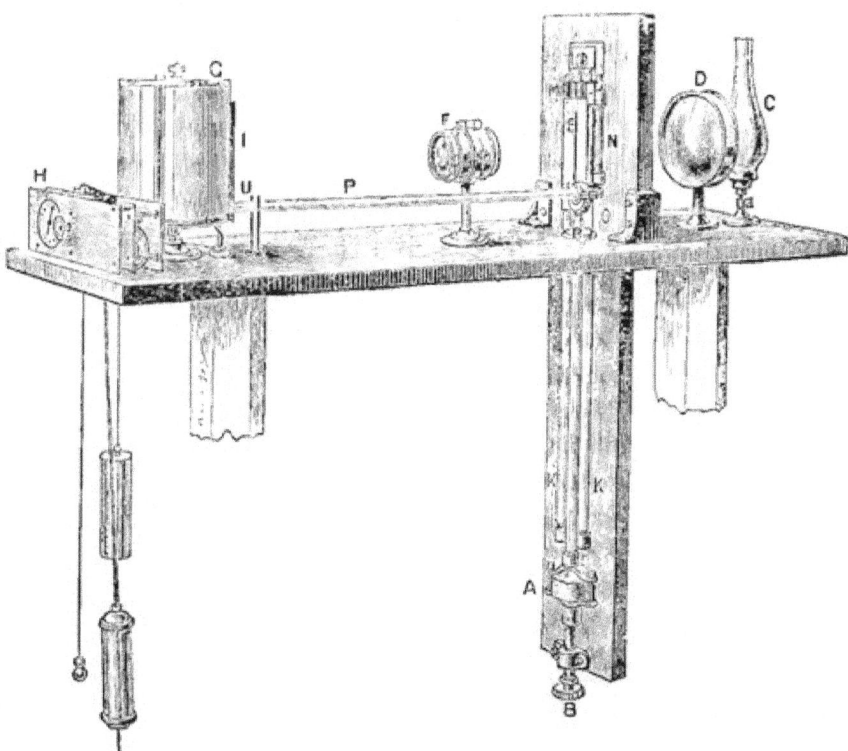

31.
Barograph.   Scale about ½.

the temperature, which will give the elevation in feet above the sea-level on reference to the table supplied by the optician from whom the instrument is purchased.

A highly-refined automatic arrangement is adopted at some observatories called a Barograph, which, by the aid of photography, becomes a self-recording mercurial barometer. It is simpler in its arrangement than the thermograph, and includes a clock of superior construction, causing a cylinder bearing photographic paper to make one complete revolution in forty-eight hours. A double combination of achromatic lenses brings to a focus rays passing through a slit placed in front of the mercurial column, behind which is a strong gaslight or paraffin lamp, the rays of which are condensed upon the slit by a combination of two plano-convex lenses.

Although a barometer is an instrument *artificially* constructed by man, it should not be forgotten that when once made the column of mercury is placed in a passive or quiescent state in direct relation with the great forces of nature, so that its indications become to some extent *natural* phenomena. This is aptly illustrated by what is called the "daily fluctuation" of the barometer which occurs in all countries, though the hours and extent vary with the latitude, diminishing as the latitude increases, according to a definite law. The phenomena does not admit of a satisfactory explanation, but is doubtless connected with the daily variations of temperature and of vapour in the air. The mercury falls *naturally* (so to speak) from nine or ten to between three and four p.m.; it then rises till between nine and ten p.m. It falls again about four a.m., and rises again about ten a.m. It is usually highest at nine a.m. and nine p.m., and lowest at three a.m. and three p.m.

These natural elevations and depressions of the mercury should be allowed for in reading the barometer, as any rise or fall in opposition to the natural rise and fall possesses for that reason increased importance. For instance, fine weather may be expected if the mercury rises between nine a.m. and three p.m.; in like manner rain may be expected should a fall take place between three p.m. and nine p.m.

It will be inferred from the preceding facts that there are certain hours better suited for "taking a reading" than others. When one observation only is made daily, noon is the best time, two observations should be made at nine a.m. and nine p.m., and for three the best hours are nine a.m. (maximum), noon (mean), and three p.m. (minimum).

The opinion generally entertained that a high barometer is an indication of fine weather, and a low one a warning of bad weather, is open to exception, and an increased value would attach to the indications of the instrument in proportion as the following points are noted and allowed for :—

1. The actual height of the mercury. 2. Whether it is rising or falling. 3. The rate of rise and fall. 4. Whether the rise or fall has been long continued.

The state of the barometer foretells *coming* weather, and when the present weather disagrees with the barometer a change will soon take place. A fall of half a tenth, or more, in an hour is a sure warning of a storm, a rapid rise is a warning of unsettled weather.

The barometer is generally lowest with wind from the S.W., and highest with wind N.E., or with a calm. N.E. and S.W. may be called the wind's poles, and the difference of height due to *direction* only from one of these bearings to another amounts to about half an inch.

TELEGRAPHIC WEATHER INTELLIGENCE.

The Meteorological Office issues (free of charge) to ports and fishing stations approved of by the Board of Trade, notices of atmospherical disturbances on or near the coasts of the British Islands.

The fact that such a notice has been received at any station is made known by a signal, which is hoisted on the receipt of the message, and remains hoisted, but only during the day-time, for the space of 48 hours.

The **Signals** are made by means of two canvas shapes, a **Cone** and a **Drum**.

The **Cone** is three feet high and three feet wide at base, and appears as a triangle when hoisted.

The **Drum** (or cylinder) is three feet high and three feet wide, and appears as a square when hoisted.

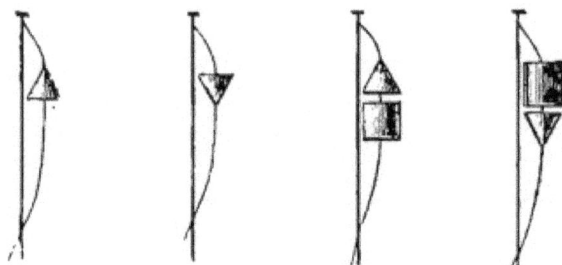

HOISTING THE SIGNALS.—The signal is kept flying until dusk, and then lowered, and hoisted again next morning; and so on until the end of 48 hours from the time the message has been issued from London (which is always marked on the telegram), unless orders are received previously to lower the signal.

At dusk, whenever a signal ought to be flying if it were daylight, a night signal may be hoisted in place of the Cone, consisting of three lanterns hung on a triangular frame, point downwards or point upwards, as the case may be. It is not considered necessary to hoist lanterns to represent the Drum. They are kept burning until late in the evening, say 9 or 10 o'clock.

No warning messages can be issued on Sundays, as the telegraph offices are almost all closed after 10 a.m. on that day, so that the signal must sometimes be kept flying on Sundays longer than is necessary. Storms, first showing signs of their approach on Sunday, will sometimes come on before Monday, so that the warning cannot be issued in time to be of service.

The hoisting of any of these signals is intended as a sign that there is an atmospherical disturbance in existence which will *probably* cause a gale from the quarter indicated by the signal used, in the neighbourhood (say within a distance of 50 miles) of the place where the signal is hoisted. Its meaning is simply, "Look out! It is probable that bad weather of such and such a character is approaching you."

Hitherto it has been found that at least *three* out of *five* signals of approaching storms (Force upwards of 8 Beaufort scale, a "Fresh Gale"), and *four* out of *five* signals of approaching strong winds (Force upwards of 6 Beaufort scale, a "Strong Breeze"), have been fully justified.

In every case some of the principal reasons which have led to the hoisting of the signal are explained in the telegram, *which is always kept posted up for public inspection while the signal is flying.*

The signal will sometimes be kept flying after the gale is over; this is the case because often one gale is followed by another before the 48 hours are out. In every case when it is thought at the Meteorological Office that immediate danger is over, orders are issued to lower the signal at once.

VERIFICATION OF INSTRUMENTS AT KEW OBSERVATORY.

The Kew Committee of the Royal Society receive, for verification and comparison with the standard instruments of the Kew Observatory, barometers, thermometers, and other instruments intended for meteorological observation or scientific investigations.

Any persons ordering instruments of opticians may direct them to be previously forwarded to the observatory for verification.

A scale of charges is issued by the Committee which is exclusive of packing and carriage, or of rail expenses, when a special messenger is sent out. The Meteorological Office, Victoria Street, London, also receives and forwards instruments for verification to the Kew Observatory.

The Committee wish it to be understood that they cannot undertake the verification of an inferior class of instruments (such as barometers mounted upon wooden frames, and thermometers not graduated on the stem), and that the superintendent of the observatory may at his discretion decline to receive such instruments as he may consider unfit for scientific observation.

BAROMETER WARNINGS.

May be expected.	
Increasing storm	If mercury falls during a high wind from S.W., S.S.W., W., or S.
Violent but short	If the fall be rapid.
Less violent but of longer continuance	If the fall be slow.
A violent storm from the N.W. or N.	If the mercury falls suddenly while the wind is due W.
N.W., N., or N.E. winds, or less wind, or less rain, or less snow	If the mercury having been at its usual height, 29·95, is steady or rising, *while the thermometer falls* and the air becomes drier.
Wind and rain from S.E., S., and S.W.	If the mercury falls, *while the thermometer rises* and the air becomes damp.
A violent storm from N.W., N., or N.E.	When the mercury falls suddenly with a W. wind.
Snow	If the mercury falls when the thermometer is low.
Less wind, or a change to N., or less wet	When the mercury rises, after having been some time below its average height.
Strong wind or heavy squalls from N.W., N., or N.E.	With the *first* rise of the mercury after it has been very low (say 29).
Improved weather	When a gradual continuous rise of the mercury occurs with a falling thermometer.

May be expected.	
Winds from S. or S.W.	If the mercury suddenly rising, the thermometer *also* rises.
Heavy gales from N.	*Soon* after the *first* rise of the mercury from a very low point.
Unsettled weather	With a *rapid* rise of the mercury.
Settled weather	With a *slow* rise of the mercury.
Very fine weather	With a continued *steadiness* of the mercury with dry air.
Stormy weather with rain (or snow)	With a rapid and considerable fall of the mercury.
Threatening, unsettled weather	With an alternate rising and falling of the mercury.
Much wind, rain, hail, or snow, with or without lightning	When the mercury falls considerably. If the thermometer be low (for the season) the wind will be N., if high, from S.
Lightning only	When the mercury is low, the storm being beyond the horizon.
Fine weather	With a rosy sky at sunset.
Wind and rain	When the sky has a sickly greenish hue.
Rain	When the clouds are of a dark Indian red.
Bad weather or much wind	When the sky is red in the morning.

EXPLANATORY CARD.
BY THE LATE VICE-ADMIRAL FITZROY, F.R.S., ETC.

WEATHER GLASSES.

THE BAROMETER RISES	THE BAROMETER FALLS
for Northerly wind	for Southerly wind
(including from North-west, by the *North*, to the Eastward),	(including from South-east, by the *South*, to the Westward),
for dry, or less wet weather,—for less wind,—or for more than one of these changes :—	for wet weather,—for stronger wind,—or for more than one of these changes :—
EXCEPT on a few occasions when rain, hail, or snow comes from the Northward with *strong* wind.	EXCEPT on a few occasions when *moderate* wind with rain (or snow) comes from the Northward.

For change of wind toward Northerly directions,—	For change of wind toward Southerly directions,—
A THERMOMETER FALLS.	A THERMOMETER RISES.

Moisture or dampness in the air (shown by a Hygrometer) increases BEFORE rain, fog, or dew.

On barometer scales the following contractions may be useful :—

RISE	FALL
FOR	FOR
NORTH	SOUTH
N.W. —N.—E.	S.E.—S.—W.
DRY	WET
OR	OR
LESS	MORE
WIND.	WIND.
EXCEPT	EXCEPT
WET FROM	WET FROM
NORTH.	NORTH.

Add one-tenth of an inch to the observed height for each hundred feet the Barometer is above the half-tide level.

The *average* height of the Barometer, in England, at the sea-level, is about 29·94 inches; and the *average* temperature of air is nearly 50 degrees (London latitude).

The Thermometer falls about one degree for each three hundred feet of elevation from the ground, but varies with wind.

"When the wind shifts against the sun,
Trust it not, for back it will run."

First rise after very low
Indicates a stronger blow.

Long foretold—long last,
Short notice—soon past.

(*In South Latitude read South for North.*)

IV.—CONDENSATION.

Dew is a deposition of moisture from the air, resulting from the condensation of the aqueous vapour of the atmosphere on substances which have become cooled by the radiation of their heat. This is, in fact, the substance of Dr. Wells's famous Theory of Dew, enunciated in 1814, and which, according to Dr. Tyndall, " has stood the test of all subsequent criticism, and is now universally accepted," and by which all the phenomena of dew may be explained.

Dr. Wells's experiments were interesting and conclusive. He exposed definite weights (10 grains) of wool to the air on clear nights, one *on* a four-legged stool, the other *under* it; the upper portion gained 14 grains in weight, the lower only 4 grains. On an evening when one portion of wool, protected by a curved pasteboard roof, gained only 2 grains, a similar portion on the top of the miniature roof gained 16 grains. A little reflection will suggest the explanation : radiation from the wool was arrested by the pasteboard cover, while the portion fully exposed to the sky lost all its heat, and thus condensation ensued. Dr. Wells speaks with such candour, and so pointedly, on this fact and its consequences, that his words may be advantageously quoted : " I had often, in the pride of half-knowledge, smiled at the means frequently employed by gardeners to protect tender plants from cold, as it appeared to me impossible that a thin mat, or any such flimsy substance, could prevent them from attaining the temperature of the atmosphere, by which alone I thought them liable to be injured. But when I had learned that bodies on the surface of the earth become during a still and serene night colder than the atmosphere, by radiating their heat to the heavens, I perceived immediately a just reason for the practice I had before deemed useless."

Familiar instances of the formation of dew will have been noted by many " watchers ; " *e. g.*, breathing on a cold pane of glass, a tumbler of cold water becoming dew-covered on being brought into a warm room, the outside of a tankard of iced claret cup, &c. When radiation is so free and rapid that the temperature is below the freezing point, the dew freezes as it forms, producing *hoar-frost*.

In our climate the air is never completely dry, nor completely saturated with moisture, and the amount of aqueous vapour held in suspension is very variable. This fact has important bearings on many branches of industry, as also on the hygienic qualities of the atmosphere. The consideration that a certain amount of moisture in the air is necessary to the continuance of health will suggest the importance of maintaining that due proportion in the atmosphere of sick rooms, where the artificial heat so injudiciously used often disturbs the healthful hygrometric condition of the air. Mr. GLAISHER is of opinion that the medical profession should enforce, as far as lies in their power, the use of this simple and effective instrument, which gives indications so important to the comfort of the patient.

The *amount* of moisture in the air is *estimated* by the use of instruments called Hygrometers, which may be thus classified:—

1. Hygrometers of Absorption.—Made with hair, catbeard, catgut, seaweed, grass, chloride of calcium.

2. Hygrometers of Condensation. — Regnault's, Daniell's, Leslie's, Dyne's.

3. Hygrometers of Evaporation.— Mason's Psychrometer, or Wet and Dry Bulb Thermometers.

By an ingenious application of the affinity of the oatbeard for moisture, Damp Detectors are constructed for tourists, commercial travellers, &c., to test moisture and avoid the consequences of sleeping in damp beds. They are strongly gilt, and resemble in size and shape a lady's watch.

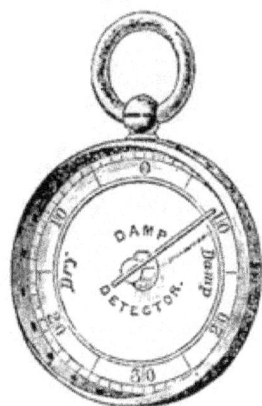

Damp Detector.
Scale about ⅔.

In Saussure's Hygrometer the

frame is of brass, and the scale of the same metal silvered. It has an attached thermometer, and the indications are the result of the contraction and expansion of a prepared human hair, consequent upon its absorbing or yielding moisture. The scale is divided on the arc of a circle, and an index needle, working on an enlarged arc, multiples the indications.

Regnault's Hygrometer (Fig. 39) consists of a thin and highly polished silver tube or bottle, into the neck of which is inserted a delicate thermometer. The bottle has a lateral tubular opening, to which is attached a flexible tube with an ivory mouthpiece. Ether is poured into the silver tube in sufficient quantity to cover the bulb of the thermometer. The ether is then agitated by breathing through the flexible tube until, by the rapid evaporation thus produced, a condensation of moisture takes place, readily observable on the

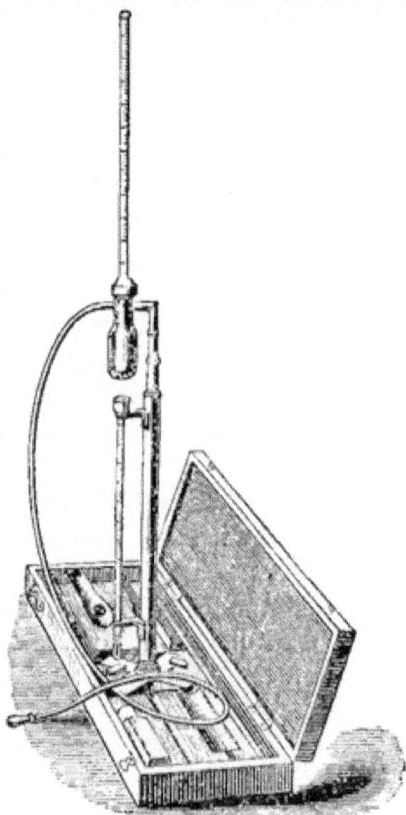

39.
Regnault's Hygrometer. Scale about $\frac{1}{10}$.

bright polished silver surface, and the temperature indicated by the thermometer at that moment is the dew-point.

Daniell's Hygrometer, or Dew-point Thermometer (Fig. 40), consists of a glass tube, bent twice at right angles, each extremity terminating in a bulb about 1½ inch in diameter, supported on a brass stand, to which a thermometer is attached to indicate the tempera-

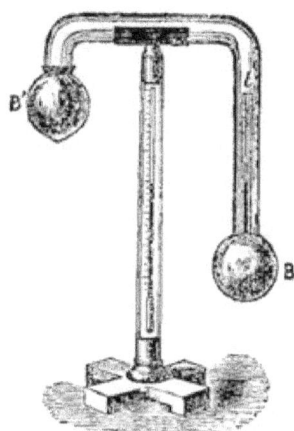

40.
Daniell's Hygrometer.
Scale about ½.

ture of the surrounding air. The lower bulb is of blackened glass, to facilitate the observation of the dew-point; it is about three parts filled with pure ether, and contains a very delicate thermometer. The upper bulb at the extremity of the short stem is transparent, but covered with thin muslin, upon which, when an observation is made, pure ether is slowly dropped. The evaporation rapidly lowers the temperature, until a moment arrives at which dew condenses on the black bulb. A quick eye is necessary to note *this* and the temperature shown by the thermometer *simultaneously*, the latter showing the degree at which the atmosphere is saturated with moisture *at the time of observation*. To avoid error, it is usual to note the temperature at which the dew disappears, and take the mean of the two temperatures.

Dyne's Hygrometer, for showing the dew-point by direct observation, by means of iced water and black glass, enables the observer to dispense with the use of ether, and shows the dew-point with great distinctness.

The hygrometer in most general use is the wet and dry bulb thermometer, and for which Mr. Glaisher has calculated an elaborate set of tables, a brief abstract of which sufficient for general purposes is subjoined.

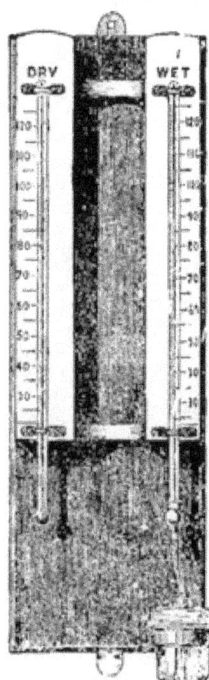

41.
Mason's Hygrometer.
Scale about ¼.

For finding the Degree of Humidity of the Air from Observations of a Dry Bulb and a Wet Bulb Thermometer, sometimes called Mason's Psychrometer.

TEMPERATURE BY THE DRY BULB THERMOMETER.	DIFFERENCE BETWEEN DRY BULB AND WET BULB READINGS.					
	2°	4°	6°	8°	10°	12°
	DEGREE OF HUMIDITY.					
34°	79	63	50	...	...	...
36	82	66	53	...	...	...
38	83	68	56	45	...	...
40	84	70	58	47	...	...
42	84	71	59	49	...	...
44	85	72	60	50	...	...
46	86	73	61	51	...	...
48	86	73	62	52	44	...
50	86	74	63	53	45	...
52	86	74	64	54	46	...
54	86	74	64	55	47	...
56	87	75	65	56	48	...
58	87	76	66	57	49	...
60	88	76	66	58	50	43
62	88	77	67	58	50	44
64	88	77	67	59	51	45
66	88	78	68	60	52	45
68	88	78	68	60	52	46
70	88	78	69	61	53	47
72	89	79	69	61	54	48
74	89	79	70	62	55	48
76	89	79	71	63	55	49
78	89	79	71	63	56	50
80	90	80	71	63	56	50
82	90	80	72	64	57	51
84	90	80	72	64	57	51
86	90	80	72	64	58	52

The total quantity of aqueous vapour which at any temperature can be diffused in the air being represented by 100, the percentage of vapour actually present will be found in the table *opposite* the temperature of the dry thermometer, and *under* the difference between the dry bulb and wet bulb temperatures. The degree of humidity for intermediate temperatures and differences

D

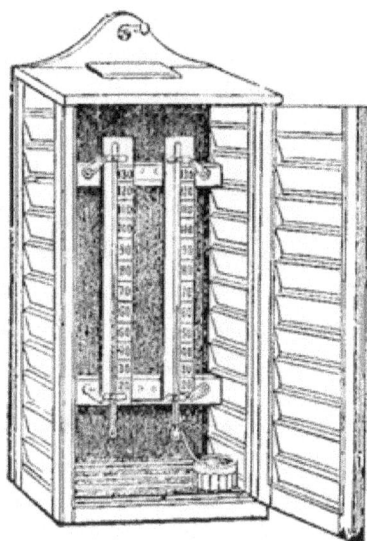

42.

Board of Trade Thermometer
Screen. Scale about ⅛.

to those given in the table can be easily estimated. Thus dry bulb 51°, wet bulb 46°, give 69 for the degree of humidity.

The instrument, as shown at page 48, consists of two thermometers attached to a support, which may be either slate or wood. The bulb of one of the thermometers has some thin muslin tied over it, and is kept *moist* by the capillary action of a thread dipping into a cistern of water placed underneath. It will be obvious that the amount of evaporation will be in proportion to the dryness of the air, and that the differences of temperature indicated by the two thermometers will be greatest when the atmosphere is dry, and least when the air is damp.

HYGROMETER PRECAUTIONS.

Hygrometers should be exposed in the shade free from air-currents.
The covering of the wet bulb must be very thin.
The supply of water must be carefully regulated.
The bulb must be constantly *moist*, yet not *too wet*.
The supply of water must be ample in dry weather.
In damp weather water must not drip from the wet bulb.
Water reservoir should be as far as possible from the dry bulb.
Dry bulb must never receive moisture from any source.
Use distilled, rain, or softest water procurable, for wet bulb.
When lime deposits from use of hard water change muslin and worsted.
Replenish reservoir after, or long before, taking an observation.
Well wash muslin and worsted before using.
Also wash occasionally while in use.
Change muslin twice a month or according to condition.
Dust and blacks must not be allowed to accumulate on muslin.
When wet bulb is frozen, wet with ice-cold water by brush.
The water will first freeze, then cool to air-temperature.
After which wet bulb falls a trifle lower than dry one.
Then temperature of evaporation may be noted.

{ In thick fog wet bulb reads *above* dry bulb.
  In cold calm, weather wet bulb reads *above* dry bulb.
  This is owing to the air being perfectly saturated.
  Covered bulb cannot therefore show temperature as well as uncovered.
  In such cases both readings are assumed to be identical.

It is important that the instrument should be protected not only from the sun's direct rays, from rain and snow, but also from wind, the currents of which would, by increasing evaporation, cause the wet bulb thermometer to indicate a temperature not strictly due to the hygrometric condition of the atmosphere. For this purpose Thermometer Screens are employed. Illustrations of two forms are shown at Figs. 42 and 43; they should be placed facing the north at a distance of four feet from the ground. Fig. 42 shows the form adopted by the Board of Trade, for marine service, while Fig. 43 shows Mr. Stevenson's double-louvred screen with perforated bottom, which ensures free ingress and egress of air, the exclusion of snow and rain, and the direct rays of the sun. Professor Wild recommends overlapping segments of sheet zinc for the construction of these screens, as possessing the advantage over wood of becoming sooner in *thermic equilibrium* with the surrounding air, and thus preventing radiation. Stevenson's Screen should be erected on legs four feet high, and should stand over grass on open

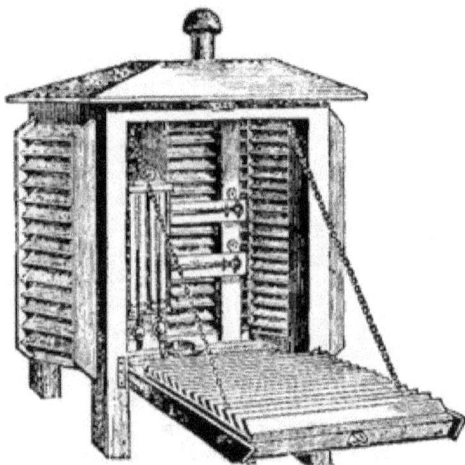

43.
Stevenson's Thermometer Screen.
Scale about $\frac{1}{10}$.

ground. It should not be under the shadow of trees, nor within twenty feet of any wall.

CLOUDS.

The important office performed by clouds in the economy of nature entitles them to extended consideration. A cloud may be defined as "water-dust," since aqueous vapour diffused through the air is invisible until the temperature is sufficiently lowered to produce condensation; no satisfactory explanation, however, has yet been given of the mode of suspension of this water-dust, nor why it remains suspended in opposition to gravitation. It is tolerably certain that electricity is not without its influence, though the apparently *stationary* character of some clouds is deceptive, for while there may be no apparent motion in the mass the particles constituting the mass are undergoing continuous renewal, which justifies the assertion of Espy that every cloud is either a forming or dissolving cloud. Aëronauts in ascending from the earth pass through many successive alternations of cloud-strata and clear air which owe their existence to the varying temperature and degrees of humidity of the atmospheric currents so superposed.

Luke Howard in his Askesian Lectures, 1802, divides clouds into three primary modifications: cumulus, stratus, and cirrus, with intermediate forms resulting from combinations of the primaries, viz., cirro-cumulus, cirro-stratus, cumulo-stratus, and cumulo-cirro-stratus or nimbus. This nomenclature is now universally adopted.

CIRRUS, or mare's tail cloud, appears as parallel, flexuous, or diverging streaks or fibres, partly straight. It is the lightest and the highest of all clouds, being seldom less than three miles, and often ten miles, above the earth, and shows the greatest variety of form. On account of its great height it is assumed to consist of minute snowflakes or crystals of ice, the refractions and reflections from which produce the halos, coronæ, and mock suns and moons which occur chiefly in this cloud and its derivatives. It retains its varied outlines

44.
Cirrus.

longer than any other cloud; at sunrise it is the first to welcome the sun's rays, and at sunset the last to part with them. It is the most useful of all clouds for weather warnings.

1. *Serene, settled weather may be expected* when groups and threads of cirri are seen during a gentle wind after severe weather.
2. *A change to wet may be expected* when, after continued fair weather, filaments, or bands of cirri (*apparently* stationary), with converging ends, travel across the sky.
3. *Rain or snow, and windy, variable weather may be expected* when cirri with fine tails vary much in a few hours.
4. *Continued wet weather may be undoubtedly expected* when horizontal sheets of cirri fall quickly and pass into the cirro-stratus.
5. *A storm of wind and rain may be expected within forty-eight hours* when fine threads of cirri seem brushed backward from the south-west.

CUMULUS.—This modification of cloud is most frequently seen on bright summer days, and is appropriately called "the day cloud" and "the summer cloud." It is formed only in the daytime, in summer calms, and results from the rise of vapours from rivers,

45,
Cumulus.

lakes, and marshes into the colder regions of the air,
the lower portions of which are readily saturable.   They
are characterized by a horizontal base, from which
they rise in dense conical and hemispherical masses
rivalling mountains in their magnitude.

Their formation is due to the convection of heat from
the earth's surface, which renders the lower atmospheric
strata capable of holding a larger amount of aqueous
vapour and simultaneously establishes an upward
current, which reaching the colder regions of the air
brings about the condensation of the aqueous vapour
into the elegant and ever-beautiful forms admired alike
"by saint, by savage, and by sage."   These begin as
mere specks, which enlarge until the sky is nearly
covered in the afternoon, and towards sunset they
generally disappear, their tops becoming cirri when the
air is dry.

1. *Fine, calm, warm weather may be expected* when
   cumuli are of moderate size and of pleasing
   form and colour.
2. *Cold, tempestuous, rainy weather may be expected*
   when cumuli cover the sky, rolling over each
   other in dense, dark, and abrupt masses.
3. *Thunder may be expected* when cumuli of hemispherical form are characterized by an extreme
   silvery whiteness.

46.
Stratus.

4. *Rain may be expected* when cumuli increase in
    number towards evening, sinking at the same
    time into the lower portions of the air.

STRATUS.—As its name implies, this is a horizontal
sheet of cloud formed near the earth at night (whence
it has been called "the night cloud") by the condensation of moist air from rivers, lakes, and marshes, or
damp ground which has lost its day-heat by radiation,
especially in calm clear evenings, after warm days. It
appears as a white mist near, and sometimes touching,
the earth. It attains its maximum density about midnight, but is dissipated by the rays of the morning sun.
Its formation, watched from a height over a large city,
is highly interesting, and is attributed by Sir John
Herschel to the soot suspended over such localities, each
particle of which acts as "an insulated radiant, collects
dew on itself, and sinks down rapidly as a heavy body."
Still more interesting is it to observe from a similar
elevation the dissipation of this cloud when the sun has
attained such an altitude that its rays fall on the upper
surface of the stratus cloud, which then heaves like the
billows of the ocean, while the whole mass seems to
rise spontaneously from the earth, and speedily vanishes
"into air, into thin air."

1. *The finest and most serene weather may be expected*
    when stratus clouds present the appearances
    just described.

CIRRO-CUMULUS, or "mackerel sky," is a well-known form of cloud occurring in small roundish masses, looking like flocks of sheep at rest, and often at great heights. It is seldom seen in winter.

1. *Increased heat may be expected* when cirro-cumuli appear.
2. *A storm or thunder may be expected* when cirro-cumuli occur mingled with cumulo-stratus in very dense, round, and close masses.
3. *Warm wet weather, and a thaw, may be expected* when cirro-cumuli occur in winter.

CIRRO-STRATUS "appears to result from the subsidence of the fibres of cirrus to a horizontal position, at the same time approaching laterally. The form and relative position when seen in the distance frequently give the idea of shoals of fish." It is called "the vane cloud" and "mackerel-backed sky."

1. *Rain, snow, and storm may be expected* when *cirro-stratus* is seen alone or mingled with cirro-cumulus, especially if the cirro-cumulus passes away.
2. *Fair weather may be expected* when from a mixture of cirro-stratus and cirro-cumulus the former disappears, leaving the latter in possession of the sky.
3. *Thunder and heat* are generally attended by waved cirro-stratus.

CUMULO-STRATUS.—This form of cloud results from the mingling of the cumulus and cirro-stratus; it appears sometimes as a thick bank of cloud with overhanging masses. The cloud known as *"distinct"* cumulo-stratus appears as a cumulus surrounded by small fleecy clouds.

1. *Thunder may be expected* when "distinct" cumulo-stratus appear.
2. *Sudden atmospheric changes may be expected* when cumulo-stratus appear.

47.
Nimbus.

NIMBUS, OR CUMULO-CIRRO-STRATUS.—The name of this cloud at once suggests that it is produced by a combination of the three primary forms of cloud. The *nimbus* is popularly known as "the rain cloud." It is really a system of clouds, having its origin chiefly in the tendency of the *cumulo-stratus* to spread, overcast the sky, and settle down to a dense horizontal black or grey sheet, above which spreads the *cirrus*, and from below which rain begins to fall.

1. *A cessation of rain may be expected* when the grey lower portion of *nimbus* begins to break up.
2. *A thunderstorm may be expected* when the *nimbus* character of the cloud is very perfect.
3. *Very copious showers may be expected* when the *cirri* projected from the top of the rain-cloud are very numerous.

AMOUNT OF CLOUDS.—Any record of the proportion of sky covered by cloud should be made on a scale of 0 to 10. A clear sky is registered 0, and a sky wholly obscured as 10, any intermediate condition being represented by 5—7, or other figures deemed appropriate by the observer. The *kind* of cloud should be noted, as also the direction in which it is driven by the wind, whether in the upper or lower strata of the air. This operation may be assisted by an ingenious arrangement, exhibited by Mr. Goddard in 1862, and called a "cloud reflector!"

obtainable at any optician's. Observations at the Greenwich Observatory establish the facts that the least amount of cloud exists during the night, especially in May and June, and the greatest amount at midday, and in winter; also that from November to February three-fourths of the heavens are obscured by sun-repelling clouds.

HEIGHT OF CLOUDS.—Great diversity of opinion exists on this point. It is asserted, on the one hand, that the region of clouds does not extend beyond five miles above sea-level, but Glaisher has attained a height of 36,960 feet, and from thence saw clouds floating at a great height above him; and it is considered probable that cirri are often ten miles above the earth.

VELOCITY OF CLOUDS.—This is of two kinds : 1st. Velocity of Propagation; and 2. Velocity of Motion. The first occurs when at a given altitude the dew-point is suddenly attained, when the sky on one occasion was covered from the eastern to the western horizon at the rate of 300 miles per hour. The second is dependent on the force of atmospheric currents, which is much greater in the upper regions of the air than in those nearer the earth. Accurate observations of the shadows of clouds, borne across the fields on a summer's day, warrant the assertion that an apparently slow motion of clouds is equal to eighty miles an hour, while a velocity of 120 miles is attained without impressing the observer with the idea of rapidity.

On the subject of clouds Admiral Fitzroy says :—

Fine weather		When clouds are "soft-looking or delicate."
Wind		When clouds are hard-edged and oily-looking.
Less wind ...		In proportion as the clouds look *softer*.
More wind ...		The harder, more "greasy," rolled, tufted, or ragged the clouds look.
Rain		When small inky-looking clouds appear.
Wind *and* rain ...	May be expected.	When light scud clouds are seen driving across heavy masses.
Wind only ...		When light scud clouds are seen alone.
Change of wind		When high upper clouds cross the sun, moon, or stars in a direction different from that of the lower clouds, or the wind then felt below.
Wind		With tawny or copper-coloured clouds.

The following " Weather Warnings " may be gathered from the COLOUR OF THE SKY :—

Whether clear or cloudy, a rosy sky at sunset presages fine weather; a sickly greenish hue, wind and rain; a red sky in the morning, bad weather, or much wind or rain; a grey sky in the morning, fine weather; a high dawn (*i. e.*, when the first indications of daylight are seen above a bank of clouds), wind ; a low dawn (*i. e.*, when the day breaks on or near the horizon), fair weather. Light, delicate, quiet tints or colours, with soft, indefinite forms of clouds, indicate and accompany fine weather; but gaudy or unusual hues, with hard, definitely outlined clouds, foretell rain and probably strong wind. Also a bright yellow sky at sunset presages wind; a pale yellow, wet; orange or copper-coloured, wind and rain : and thus, by the prevalence of red, yellow, green, grey, or other tints, the coming weather may be told very nearly ; indeed, if aided by instruments, almost exactly.

After fine, clear weather the first signs in a sky of a coming change are usually light streaks, curls, wisps, or mottled patches of white distant cloud, which increase and are followed by an overcasting of murky vapour that grows into cloudiness. This appearance, more or less oily or watery as wind or rain will prevail, is an infallible sign.

Usually, the higher and more distant such clouds seem to be, the more gradual, but general, the coming change of weather will prove.

Misty clouds, forming or hanging on heights, show wind and rain coming, if they remain, increase, or descend ; if they rise or disperse, the weather will improve or become fine.

Fine weather ...		When the sky is grey in the morning.
Wind		With a high dawn.
Fair weather ...		With a low dawn.
Wind	May be expected	When the sky at sunset is of a *bright* yellow.
Rain		When the sky at sunset is of a *pale* yellow.
Wind and rain ...		When the sky is orange or copper colour.
Fine weather ...		When the sky has light, delicate, quiet tints and soft, indefinite forms of clouds.
Rain and wind ...		When the sky has gaudy, unusual hues, with hard, definitely outlined clouds.

Fair weather ...		When sea-birds fly out early and far to seaward.
Stormy weather		When sea-birds hang about the land, or fly inland.
Fair weather ...		When dew is deposited. Its formation never *begins* under an overcast sky, or when there is much wind.
Rain		On what is called a good *hearing* day.
Rain		When remarkable clearness of atmosphere, especially near the horizon, exists, distant objects, such as hills, being unusually visible or well defined.

May be expected.

RAIN.

The atmosphere at a given temperature is capable of retaining only a given quantity of aqueous vapour, invisibly diffused through it, at which temperature it is said to be *saturated*. Should the temperature from any cause be lowered, the aqueous vapour at once becomes visible in the form of either cloud, dew, rain, snow, or hail. It has already been shown that, although marshes and rivers, inland seas and lakes, yield by evaporation watery vapours to the air, the ocean is the great source of rain, whence it is lifted in vast quantities by the sun's radiant heat, to be subsequently condensed by passing into cooler regions, or by contact with cold mountain peaks, falling to earth as a fertilizing shower or a devastating flood.

Sir John Herschel accounts for the formation of raindrops by saying:—"In whatever part of a cloud the original ascensional movement of the vapour ceases, the elementary globules of which it consists being abandoned to the action of gravity, begin to fall. The larger globules fall fastest, and if (as must happen) they overtake the slower ones, they incorporate, and the diameter being thereby increased, the descent grows more rapid, and the encounters more frequent, till at length the globule emerges from the lower surface of the cloud at the 'vapour plane' as a drop of rain, the size of the drops depending on the thickness of the cloud stratum and its density."

Rain is very unequally distributed, there being por-

tions of the torrid zone where it *never* falls, one locality
in Norway where it falls three days out of four, and
another on the western side of Patagonia, at the base of
the Andes, where it falls every day. The quantities recorded as having fallen at one time in some localities are
simply appalling. A fall of one inch is considered a very
heavy rain in Great Britain, and this fact will enable
the reader partially to realize the following stupendous
recorded falls :—Loch Awe, Scotland, 7 inches in 30
hours ; Joyeuse, France, 31 inches in 22 hours ; Gib-
raltar, 33 inches in 26 hours; hills above Bombay, 24
inches in one night ; and on the Khasia Hills, where the
annual rainfall is 600 inches, 30 inches have been known
to fall on each of five successive days. Mr. G. J. Symons,
the able editor of the "Meteorological Magazine," and in-
defatigable superintendent of 2,000 Rain Gauges throughout the United Kingdom, has compiled a table, showing
the equivalents of rain in inches, its weight per acre,
and bulk in gallons, the following portion of which,
while very useful to the farmer, will enable the curious
reader to make some interesting calculations, based on
the figures quoted above :—

TABLE SHOWING EQUIVALENT OF INCHES OF RAIN IN GALLONS,
AND WEIGHT PER ACRE.

Inches of Rain	0·1	0·2	0·3	0·4	0·5	0·6	0·7	0·8	0·9	1 ·in.
Tons per Acre	10	20	30	40	50	61	71	81	91	101
Gallons per Acre	2262	4525	6787	9049	11312	13574	15836	18098	20361	22623

The instruments called Rain Gauges or Pluviometers
are, as their name implies, constructed to measure the
amount of rain falling in any given locality, and those
in most general use have this principle in common : that
the graduated glass always bears a definite relation to the
area of the receiving surface. A very extraordinary and
hitherto unexplained fact in connection with the fall of
rain, and which justifies the opinion that its formation is

not limited to the region of visible cloud, is that a series of rain gauges placed at different elevations above the soil are found to collect very different quantities of rain, the amount being *greater* at the *lower* level. Thus, twelve months' observations by Dr. Heberden determined that the amount of rain on the top of Westminster Abbey was only twelve inches, that on a house close by but much lower eighteen inches, and on the ground during the same interval of time twenty-two inches. Accordingly, ten inches is the height at which meteorologists have agreed the edge of the rain gauge should be placed from the ground. The spot chosen should be perfectly level, and at least as far distant from any building or tree as the building or tree is high, and, if the gauge cannot be equally exposed to all points, a south-west aspect is preferable. It is also important that the rain gauge should be well supported, in order to avoid its being blown over by the wind; and, should frost follow a fall of rain, the instrument should be conveyed to a warm room to thaw before measuring the collected contents. The graduated glass furnished with each instrument should stand quite level when measuring the rain, and the reading be taken midway between the two apparent surfaces of the water.

The best form of rain gauge is that in use in the Meteorological Office.

Howard's Rain Gauge consists of a vertical glass receiver, or bottle, through the neck of which the long terminal tube of a circular funnel, five inches in diameter, is inserted. A metal collar or tube fits over the outside of the

43.
Howard's Rain Gauge.
Scale about ⅕.

neck of the receiver, and aids in keeping the funnel level, while the tube extends to within half an inch of the bottom, thus ensuring the retention of every drop of rain which falls within the area of the funnel. The glass vessel furnished with the instrument is graduated to 100ths of an inch. A modification of this instrument is made with a glass tube at the side graduated to inches, 10ths, and 100ths, showing the amount of rainfall by direct observation, thus dispensing with the use of a supplementary graduated measure.

In Glashier's Rain Gauge special provision is made, in two ways, to prevent possible loss by evaporation, even in the warmest months of the year. 1. The receiving vessel is partly sunk beneath the soil, thus keeping the contents cool. 2. The receiving surface of the funnel, accurately turned to a diameter of eight inches, terminates at its lower extremity in a curved tube, which, by always retaining the last few drops of rain, prevents evaporation. The graduated vessel, in this instance also, is divided to 100ths of an inch, having due regard to the larger area, 8 in. of the funnel. For use in tropical climates, where, as has been shown, the rainfall is excessive, a modification of this instrument is supplied by the instrument makers, having an extra large receiver and tap for drawing off the collected rain.

Luke Howard, in his " Climate of London," says : " It must be a subject of great satisfaction and confidence to the husbandman to know at the beginning of a summer, by the certain evidence of meteorological results on record, that the season, in the ordinary course of things, may be expected to be a dry and warm one, or to find, in a certain period of it, that the average quantity of rain to be expected for the month has fallen. On the other hand, when there is reason, from the same source of information, to expect much rain, the man who has courage to begin his operations under an unfavourable sky, but with good ground to conclude, from the state of his instruments and his collateral knowledge, that a fair interval is

approaching, may often be profiting by his observations, while his cautious neighbour, who 'waited for the weather to settle,' may find that he has let the opportunity go by." This superiority, however, is attainable by a very moderate share of application to the subject, and by the keeping of a plain diary of the barometer and rain gauge, with the hygrometer and vane under his daily notice.

Symons's Rain Gauge re-

49.

Symons's Rain Gauge.
Scale about ¼.

sembles Howard's, but has the advantage of having the glass receiver enclosed in a black or white japanned metal or copper jacket with openings permitting an approximate observation of the collected rain. The metal jacket is also furnished with strong iron spikes, which are firmly pressed into the soil, as shown at Fig. 49, thus ensuring perfect steadiness by its power to resist the wind. The graduated measure contains half an inch of rain (for a 5 inch circle) divided into 100ths.

Mr. Symons has devised another rain gauge of so ingenious and interesting a character that it needs only to become generally known among amateur meteorologists to be in universal demand. By its means an observer at a distant

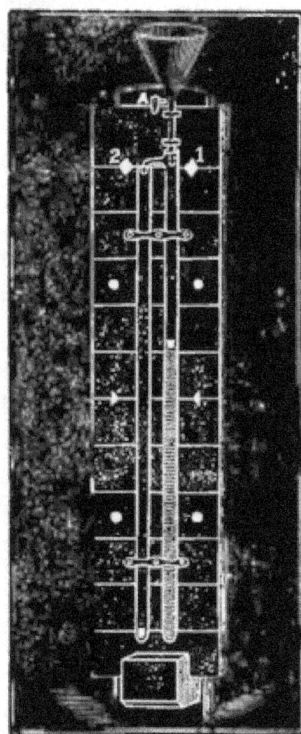

50.
Symons's Storm Rain Gauge.
Scale about 1/12.

window may read off the rain as it falls. It is shown at Fig. 50, where the usual 5-inch funnel surmounts a long glass tube attached to a black board bearing a very open scale marking tenths of an inch in *white* lines ; a white float inside the tube constitutes the index, which rises as the rain increases in quantity. If, as sometimes happens during a thunderstorm, the rainfall is excessive, a second tube on the left permits the measurement of a second inch of rain. It will be obvious that if the *time* at which the rain begins to fall be noted the *rate* at which it falls, as well as the quantity, is indicated at sight by this instrument.

Crossley's Registering Rain Gauge has a receiving

51.
Beckley's Pluviograph.   Scale about ½.

surface of 100 square inches.   The rain falling within this area passes through a tube to a vibrating bucket, which sets in motion a train of wheels, and these move the indices on three dials, recording the amount of rain in inches, 10ths, and 100ths.

E

Printed directions are furnished with each instrument, and the simplicity of the mechanism ensures due accuracy. A test measure, holding exactly five cubic inches of water, sent with each gauge, affords the means of checking its readings from time to time.

Beckley's Pluviograph possesses the exceptional merit of recording with equal precision all rainfalls, from a slight summer shower to a heavy storm of rain. It may be

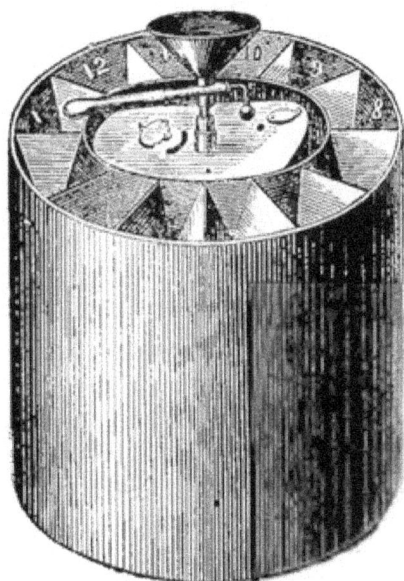

52.   53.

Stutter's Self-recording Rain Gauge. Scale about ⅓.

placed in a hole in the ground, with the receiving surface raised the standard height of ten inches above its level.

Fig. 51 illustrates the construction of the instrument.

The funnel has a receiving surface of 100 square inches, protected by a lip 1¼ inch deep, to retain the splashes. The rain flows into a copper receiving vessel on the right, which, floating in a cistern of mercury, sinks and draws down with it a pencil, which records the event on a white porcelain cylinder moved by a clock. When the receiving vessel is full the syphon comes into action, rapidly drawing off *the whole* of the water, the vessel rising almost at a bound, the action being recorded by a vertical line on the porcelain cylinder. Two or more cylinders are supplied with each instrument; and, as the pencil marks are readily removed by a little soap and water, a clean one may be always kept at hand for exchange once in every twenty-four hours.

The Rev. E. Stutter's Self-recording Rain Gauge is ingenious, and for a self-recording instrument is very moderate in price, while it efficiently shows the rainfall for every hour in the twenty-four (Figs. 52, 53).

An eight-day clock with its upright spindle revolves a small funnel with a sloping tube, the end of which passes successively over the mouth of the twelve or twenty-four compartments in the rim of the instrument; beneath each compartment is placed a tube, as shown in the sectional figure. All rain received by the outer funnel drips into the smaller revolving funnel, and flows down the sloping tube, the end of which is timed to take an hour in passing over each compartment, so that the rain, for example, which falls between twelve and one o'clock will be found in the tube marked 1. Each tube can contain half an inch of rain, and any overflow falls into a vessel beneath, and can be measured; the tube which has overflown shows the hour.

V.—MOTION.

Wind is air in motion. The motion of the air is caused by inequality of temperature. The earth becomes warmed by the sun, and radiates the heat thus acquired back upon the air, which, expanding and becoming lighter, ascends to higher regions, while colder and denser currents rush in to occupy the vacated space. Two points are to be noted in connection with this rush of air which we call wind, viz., its *direction* and *velocity* or *force*. Both are estimated scientifically by instruments called Anemometers,* while mariners and the dwellers on our coasts have a nomenclature of their own by which to indicate variation in the *force* of the wind,

* *Anemos*, the wind; *metron*, measure.

founded on the amount of sail a vessel can carry with safety at the time.   In the matter of *direction* winds are classed as constant, periodical, and variable.

CONSTANT WINDS.—*The Trade Winds.*—The violent contrast between the temperature of the equator and the poles is well known, and from the vast area included within the tropics ascending currents of rarefied air are incessantly rising and being as incessantly replaced by a rush of cold air from the poles to the equator.   Were the earth stationary, this interchange would be of the simplest kind; on arriving within the influence of the ascending equatorial current the air from the poles would be carried to the higher regions and turning over would proceed to the poles, and, becoming cold and dense in traversing the higher stratum, would descend and resume its course *ad infinitum.*   The revolution of the earth on its axis changes all this : the first effect is that the air at the equator is borne along with the earth at the rate of seventeen miles a minute from west to east, a rate which diminishes at 60° of latitude to one-half that velocity, until at the poles it is nothing; consequently a *slow* north wind flowing to the equator is continually passing over places possessing a higher velocity than itself, and not immediately acquiring that velocity, there is according to the law of the composition of forces a compromise effected resulting in a north-east wind.   In a similar manner the same process in the southern hemisphere results in a south-east wind.   These winds have acquired the name of *Trade Winds* on account of the facilities afforded to navigation by their constancy. The *North Trades* occur in the Atlantic between 9° and 30° and in the Pacific between 9° and 26°.   The *South Trades* occur in the Atlantic between lat. 4° N. and 22° S. and in the Pacific between latitude 4° N. and 23½° S. These limits extend northward with the sun from January to June, and southward from July to December.

Parallel to the equator and extending between 2° and 3° on each side is a broad belt, where the north and south trades neutralize each other, producing what is called the "*Region or Belt of Calms.*" Though wind is absent, thunderstorms and heavy rains are of daily occurrence.

When Humboldt ascended Teneriffe the trade wind was blowing at its base in the usual direction, but on arriving at the summit he found a strong wind blowing in the opposite direction. Observation has shown that this upper current prevails north and south of the equator, and that, after passing the limit of the trade winds, it descends to form the south-*west* winds of the north temperate zone and the north-*west* winds of the south temperate zone; the *westing* being due to the same cause as the *easting* in the regular trades, viz., the rotation of the earth on its axis. These winds are called the Return Trades, but are not equal in constancy to the regular trade winds.

PERIODICAL WINDS.—*Land and Sea Breezes* occur on the coasts, chiefly in tropical countries, but sometimes in Great Britain during the summer months when the land during the day becomes very hot, causing an ascending column of air, which is replaced by a comparatively colder stream flowing inwards from the sea. At sunset the conditions are reversed, the earth becomes rapidly cooled by radiation, the sea continuing comparatively warm, the air over it ascends, and its place is supplied by a cold breeze, which "blows off the shore," as illustrated by the diagrams and the following experiment:— In the centre of a large tub of water float a water plate containing hot water; imagine the former to be the ocean and the latter the heated land, rarefying the air over it. Light a candle and blow it out and hold it while still smoking over the cold water, when the smoke will be seen to move towards the plate. The reverse of this takes place if the tub be filled with hot water and the water plate with cold. When this phenomenon takes place on a large scale, as in the case of the north trade winds being drawn from their course by the heated shores of Southern Asia, the gigantic sea breeze thus produced is called the south-west monsoon. This occurs from April to October, when the sun is north of the equator. When the sun is south of the equator—that is, from October to April—the analogue of the land breeze is produced, and is called the north-east monsoon.

VARIABLE WINDS.—The character of this class of winds is determined by the physical configuration of the country in which they occur. Some tracts are marked by luxuriant vegetation, others are bare. Here mountains lift their awful fronts and "midway leave the storm," there an arid plain extends itself to the sea-shore, or inland, towards a chain of lakes. Within the tropics these purely local conditions are insufficient to overcome the force of the prevalent atmospheric currents: such, however, is not the case beyond the tropical zone. There the variable winds prevail, for which space permits only the mention of their names:—The *Simoom* (from the Arabic *samma*, hot), peculiar to the hot sandy deserts of Africa and Western Asia. The *Sirocco* blows over the two Sicilies as a hot wind from the south. It extends sometimes to the shores of the Black and Caspian seas, spreading death among animals and plants. The *Solano* prevails at certain seasons in the south of Spain: its direction is south-east. The *Harmattan* is another wind of the same class, peculiar to Senegambia and Guinea. The *Puna Winds* blow for four months over a barren tableland called the Puna, in Peru. They are a portion of the south-east trade winds, which, having crossed the Pampas, are thereby deprived of moisture, and become the most parching wind in the world. The *East Winds*, peculiar to the spring in Britain, blowing as they do through Russia, over Europe, are a portion of the great polar current, distinctive of that season of the year. They are dry and parching, every one being familiar with the unpleasant bodily sensations attendant on this much-abused and yet most beneficent wind.

The *Etesian Winds* are drawn from the north across the Mediterranean by the great heat of the African desert. The *Mistral* is a strong north-west wind peculiar to the south-east of France. The *Pampero* is a north-west wind, blowing in summer from the Pampas of Buenos Ayres.

As long ago as the year 1600 Lord Bacon remarked that the preponderating tendency of the wind was decidedly to veer *with* the sun's motion, thus passing from N. through N.E., E., S.E., to south, thence through S.W., W.N.W., to N. ; also, that it often makes a complete circuit in that direction, or more than one in succession (occupying sometimes many days in so doing), but that it rarely backs, and very rarely or never makes a complete circuit in the contrary direction. The merit of having first demonstrated that this tendeucy is a direct consequence of the earth's rotation is due to Professor Dove, of Berlin, who has also shown that the three systems of atmospheric currents just treated of, viz., the constant, periodical, and variable winds, are all amenable to the same influence.

As to the *mode* of observing the wind, Admiral Fitzroy recommends that a true east and west line should be marked *about the time of the equinox,* and the north, south, and other points of the compass being added, to take the bearings of the wind in relation to a dial so prepared, the indications of the *lower* stratum of clouds in conjunction with vanes and smoke being preferred to any other.

The direction of the wind should always be given according to *true,* and not to *compass bearings.* Two points to the westward nearly represents the amount of " Variation of the Compass" for the British Isles, which yields the following table for the conversion of directions observed by the compass in Great Britain and Ireland to approximate true bearings.

Compass bearings.	N	NNE	NE	ENE	E	ESE	SE	SSE
True bearings.	NNW	N	NNE	NE	ENE	E	ESE	SE
Compass bearings.	S	SSW	SW	WSW	W	WNW	NW	NNW
True bearings.	SSE	S	SSW	SW	WSW	W	WNW	NW

"One may call a very simple diagram, a circle divided by a diameter from north-east to south-west, the *thermometer compass.* While the wind is shifting from south-west, by west, north-west, and *north to north-east,* the thermometer is falling, but while shifting from north-east, by east, south-east and south, towards *south and south-west,* the thermometer is rising. Now the barometric column does just the reverse. From north-east the barometer falls as the wind shifts through the east to south-east, south, and south-west, and from the south-west, as the wind shifts round northward to north-east, the barometer rises—it rises to west, north-west, north, and north-east.

"The effect of the wind thus shifting round when traced upon paper by a curve, seems certainly wave-like to the eye ; but I believe it to be simply consequent on the wind shifting round the compass, and indicating alteration in the barometric column.

"If the wind remained north-east, say three weeks, there would be no wave at all—there would be almost a straight line along a diagram (varying only a little for *strength*). The atmospheric line, in such a case, remains at the same height, and the barometer remains at 30 inches and (say) some three or four-tenths, for weeks together. So likewise when the wind is south-westerly a long time, or near that point, the atmospheric line remains *low,* towards 29 inches. Thus, such 'atmospheric waves' may be an optical delusion.

"The diagram alluded to above shows how the barometer and thermometer may be used in connection with each other in foretelling wind, and consequently weather, that is coming on, because *as the one rises, the other* generally *falls,* and if you take the two together and confront with their indications the amount of moisture in the air at any time, you will scarcely be mistaken in knowing what kind of weather you are likely to have for the *next two or three days,* which for the gardener, the farmer, soldier, sailor, and traveller must be frequently of considerable importance." *

We are indebted to M. Buys Ballot, a Dutch meteorologist, for an invaluable generalization, the importance of which it is almost impossible to over-estimate. This distinguished *savant* says :—"It is a fact above all doubt that the wind that comes is nearly at right angles to the line between the places of highest and lowest barometer readings. The wind has the place of lowest barometer at its left hand, and is stronger in proportion as the difference of barometer readings is greater." These facts have been variously stated by other writers ; for example : " Stand with your back to the wind, and the barometer will be lower on your left hand than on your right ;" " Facing the wind the centre of depression bears in the right-hand direction," state-

* The late Admiral Fitzroy.

ments which can be verified at any time by a brief
study of the "Weather Charts" now published in the
daily journals. The value of the law consists in its
connecting the surface winds of our planet with the
actual pressure of the air itself, and it admits of the
following tabulation :—

The wind is NORTHERLY when the BAROMETER is, in the	The wind is SOUTHERLY when the BAROMETER is, in the	The wind is EASTERLY when the BAROMETER is, in the	The wind is WESTERLY when the BAROMETER is, in the
N. & S. } about equal. E. Low. W. High.	N. & S. } about equal. E. High. W. Low.	N. High. S. Low. E. & W. } about equal.	N. Low. S. High. E. & W. } about equal.

which can be verified by the reader from the daily
Weather Charts in the newspapers.

The above are deductions from Buys Ballot's Law,
still further impressed on the memory by taking four
outline maps of the British Isles, inserting the names of
Thurso, Penzance, Yarmouth, and Valentia, with barometer readings of the kind above named at each place,
and then drawing a large arrow in red ink across the
centre of each map in the direction appropriate to the
readings.

Mr. Strachan, in his able pamphlet on "Weather
Forecasts," puts the matter thus : "It follows from
Ballot's Law that in the northern temperate zone the
winds will circulate around an area of low atmospherical
pressure in the *reverse direction* to the movement of the
hands of a watch, and that the air will flow away from
a region of high pressure, and cause an apparent circu-
lation of the winds around it, *in the direction* of watch
hands." And as the result of a careful digest of data
contained in the eleventh number of meteorological
papers, published by the Board of Trade, he has
established the following valuable propositions. As
introductory to the propositions, it should be stated that
the positions of observations were the following :—

Places.			Latitude.	Longitude.
Nairn	...	...	57° 29′ N.	4° 13′ W.
Brest	...	...	48° 28′	4° 29′ W.
Valentia	...	...	51° 56′	10° 19′ W.
Yarmouth	...	...	52° 37′	1° 44′ E.
Portrush (or Green-				
castle)	...	...	55° 12′	6° 40′ W.
Shields	...	...	55° 0′	1° 27′ W.

Nairn and Brest are situated nearly on the same
meridian, about 540 geographical miles apart. Valentia
and Yarmouth are nearly on the same parallel of latitude, about 450 miles apart. Portrush and Shields,
distant 180 miles, are on a parallel which is nearly as
remote from the parallel of Nairn as that of Valentia
and Yarmouth is from the one passing through Brest ;
and Shields is about as much to the westward of
Yarmouth as Portrush is to the eastward of Valentia.
When observations have not been obtainable for Brest,
those made at Penzance have been used instead.

Proposition 1.—Whenever the atmospherical pressure
is greater at Brest than at Nairn, while it is of the same
or nearly the same value at Valentia and Yarmouth,
being gradually less from south to north, the winds over
the British Isles are *westerly.*

Proposition 2.—Whenever the pressure at Nairn is
greater than at Brest, while its values at Valentia
and Yarmouth are equal, or nearly so, the winds
over the British Isles are *easterly.*

Proposition 3.—Whenever the pressure at Valentia is
greater than at Yarmouth, while its values at Brest
and Nairn are nearly equal, the winds over the British
Isles are *northerly.*

Proposition 4.—Whenever the pressure at Yarmouth
exceeds that at Valentia, while there is equality of
pressure at Nairn and Brest, the winds of the British
Isles are *southerly.*

Proposition 5.—Whenever the pressure of the atmosphere is equal, or nearly so, at Brest, Valentia, Nairn,
and Yarmouth, and generally uniform, the winds over

the British Isles are variable in direction and light in force.

The data from which the foregoing propositions were deduced, and indeed all other cases calculated by Mr. Strachan, show in every well-marked instance that when the atmospherical pressure was

(1) greater in the south than in the north, the wind had westing;

(2) greater in the north than in the south, the wind had easting;

(3) greater in the east than in the west, the wind had southing;

(4) greater in the west than in the east, the wind had northing;

(5) uniformly high, or uniformly low, variable light winds (with fine weather in the former case, and vapoury or wet weather in the latter).

Conditions (1) and (3) give winds from the S.W. quarter.

Conditions (1) and (4) give winds from the N.W. quarter.

Conditions (2) and (4) give winds from the N.E. quarter.

Conditions (2) and (3) give winds from the S.E. quarter.

These principles may be employed to set forth the mode of foretelling the impending change of wind as regards its direction and force; for the atmospherical pressure may change—

(a) uniformly over the whole area of observation;

(b) by increasing in the south, or (which causes a similar statical force) by decreasing in the north;

(c) by increasing in the north, or (which has the same effect) by decreasing in the south;

(d) by increasing in the west, or (which has the same effect) by decreasing in the east;

(e) by increasing in the east, or (which has the same effect) by decreasing in the west;

Scale, 0 to 6.	Pressure in pounds per square foot.	Miles per hour.	Seaman's Nomenclature.	Scale, 0 to 12.	Beaufort Scale.
0·0	0·00	2	Calm ...	0	
0·5	0·25	5	Light Air ...	1	Just sufficient to make steerage way.
1·0	1·0	10	Light Breeze ...	2	With which a ship with all sail set would go in smooth water. — 1 to 2 knots.
1·5	2·25	15	Gentle Breeze ..	3	3 to 4 ,,
2·0	4·0	20	Moderate Breeze	4	5 to 6 ,,
2·5	6·25	27	Fresh Breeze ...	5	Royals, &c. (In which she could just carry)
3·0	9·0	35	Strong Breeze ...	6	Single Reefs and T.G. Sails.
		42	...		Double Reefs and Jib, &c.
3·5		50	Moderate Gale ...	7	Triple Reefs, &c.
4·0	16·00	60	Fresh Gale ...	8	Close Reefs and Courses.
4·5	20·25		Strong Gale ...	9	In which she could just bear close-reefed
5·0	25·00	70	Whole Gale ...	10	Maintopsail and reefed Foresail.
5·5	30·25	80	Storm ...	11	Under Storm Staysails or Trysails.
6·0	36·00	90	Hurricane ...	12	Bare Poles.

With (*a*) similar wind and weather will continue.

   „   (*b*) winds will veer towards west.

   „   (*c*)   „   „   east.

   „   (*d*)   „   „   north.

   „   (*e*)   „   „   south.

"The probable strength of wind will be in proportion to the rate of increase of statical force, or differences of barometrical readings. The position of least pressure must be carefully considered; as, in accordance with the law, the wind will blow around that locality. The same remark applies to areas of high pressure, which, however, very rarely occur in a well-defined manner over the British Isles."

Referring to the table on page 76, the scale 0 to 6 was formerly used by meteorological observers at land stations, and it was intended to express, when the square of the grade was obtained, the pressure of the wind as given in the second column.

"The velocity is an approximation as near as can be obtained, from the values assigned by Neumayer, Stow, Laughton, Scott, Harris, James, &c." *

Few meteorological axioms are better established than that which embodies the fact that "every wind brings its weather," and the primary cause of wind being the motion of the air induced by rarefaction, it is obvious that there is a constant tendency for the equatorial and polar currents in any locality to establish an equilibrium, and this consideration is found to facilitate weather predictions for extended periods. Thus, in consequence of the unusual prevalence of *east* winds in the spring of 1862, a wet summer was predicted. The prediction was fully borne out by an incessant continuance of *south-west winds*, with clouded skies and the usual accompaniment of deluges of rain. These winds continuing, with slight intermissions only, till the spring of the following year, less than the usual number of south-west winds was looked for during the

* Strachan's "Portable Meteorological Register," 4th edition.

summer; the result fully justified the anticipation, the summer of 1863 being fine and warm, especially during the earlier portion. Similarly, the summer of 1877 was a dry and cool one, thus establishing an equilibrium after the long continuance of warm and wet months in the winter of 1876-7.

The scientific research and mechanical ingenuity directed of late years to producing trustworthy estimates of the direction, pressure, and velocity of the wind, have resulted in the production of a series of instruments, possessing great precision and accuracy.

The *direction* of the wind is indicated by vanes, a very efficient form of which is shown at Fig. 54, the *velocity* by revolving cups, and the *pressure* by the pressure plate and by calculation from the known velocity.

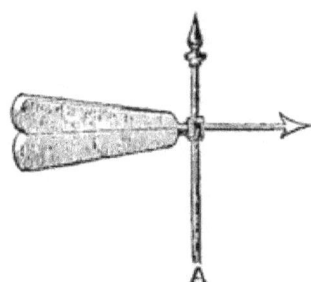

The Pendulum Anemometer (Fig. 56) shows in a simple manner the direction and pressure of the wind. The peculiarly shaped vane ensures the surface of the swinging pressure plate B being always kept towards the wind. The pendulum plate hangs, during a calm, quite vertically, indicating zero, and as the pressure increases it will be raised through all degrees of elevation from 1 to 12. The vane is perforated with holes large enough to be visible at some distance from the ground, the 5 and 10 being specially larger, so that the angle to which the pressure plate is raised can be quickly noted.

54.
Wind Vane. Scale about ⅟₇₅.

55.
Compass Bearings. Scale about ⅟₂₅.

There is a simple contrivance (for the convenience of travellers) called a Portable Wind Vane, or Anemometer. It is furnished with a compass and bar needle, &c., and will tell the true direction of the wind to within a half point.

Lind's Anemometer or Wind Gauge ranks among the earliest forms of instruments designed to estimate the force of the wind. It consists of a glass syphon, the limbs of which are parallel to each other, mounted on a vertical rod, on which it freely oscillates by the action of the vane which surmounts it. The upper end of one limb of the syphon is bent outward at right angles to the main direction, and the action of the vane keeps this open end of the tube always towards the quarter from whence the wind blows. Between the limbs of the syphon is placed a scale graduated from 0 to 3 in inches and 10ths, the zero being in the centre

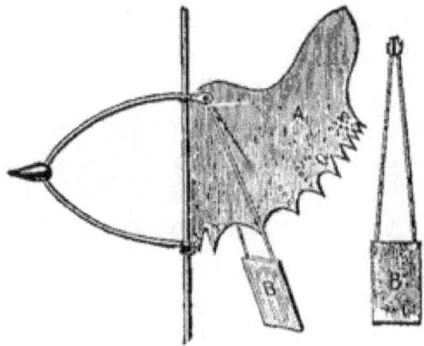

Prestel's Pendulum Anemometer.
Scale about $\frac{1}{12}$.

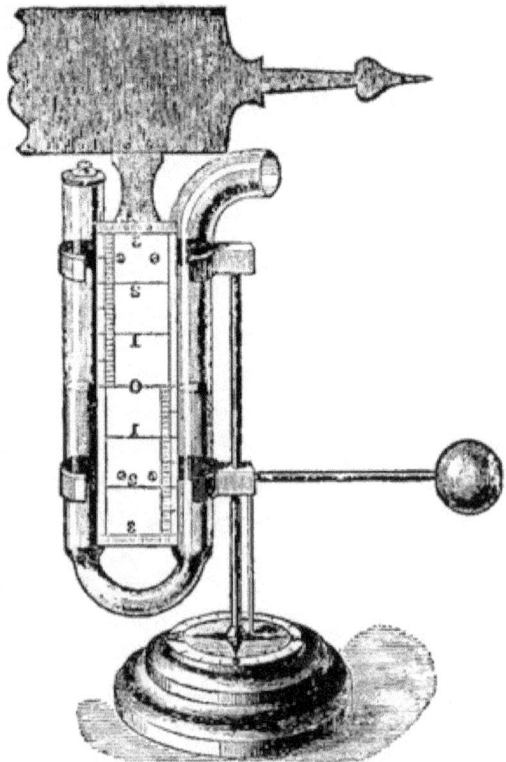

Lind's Anemometer. Scale about $\frac{1}{4}$.

of the scale. When the instrument is used, it is only necessary to fill the tube with water to the zero of the scale, and then expose it to the wind. The natural consequence of wind acting on the surface of the water is to depress it in one limb and raise it in the other, and the sum of the depression and elevation is the height of a column of water which the wind is capable of sustaining at the time of observation. Sudden gusts of wind are apt to produce a jumping effect on the water in the tube, and to diminish this the bend of the syphon is contracted. A brass plate is attached to the foot of the instrument, bearing the letters indicating the cardinal points of the compass, to show the direction of the wind.

Dr. Robinson, of Armagh, introduced an instrument, in 1850, which consists of four hemispherical copper cups attached to the arms of a metal cross. The vertical axis upon which these are secured has at its lower extremity an endless screw placed in gear with a train of wheels and pinions. Each wheel is graduated respectively to 1-10th, 1 mile, 10 miles, 100 miles, 1,000 miles, and these revolve behind a fixed index, the readings of which are taken according to the indications on the dials.

Dr. Robinson entertained the theory that the cups (measuring from their centres) revolved with one-third of the wind's velocity; and this theory having been fully supported by experiment, due allowance has been made in graduating the wheels so that the true velocity is obtained by direct observation.

In an improved form of this anemometer the hemispherical cups are retained, but the index portion of the instrument consists of two graduated concentric circles, the inner one representing five miles divided into 10ths, and the outer one bearing 100 divisions, each of which is equivalent to five miles. At the top of the dial is a fixed index, which, as the toothed wheel revolves, marks on the inner circle the miles (up to five) and 10ths of miles the wind has travelled, while a movable index, which revolves with the wheel, indicates on the outer circle the passage of every five miles.

This instrument can be made very portable by removing the arms bearing the cups, when the whole may be packed with iron shaft in a case 15 × 13 × 4 inches.

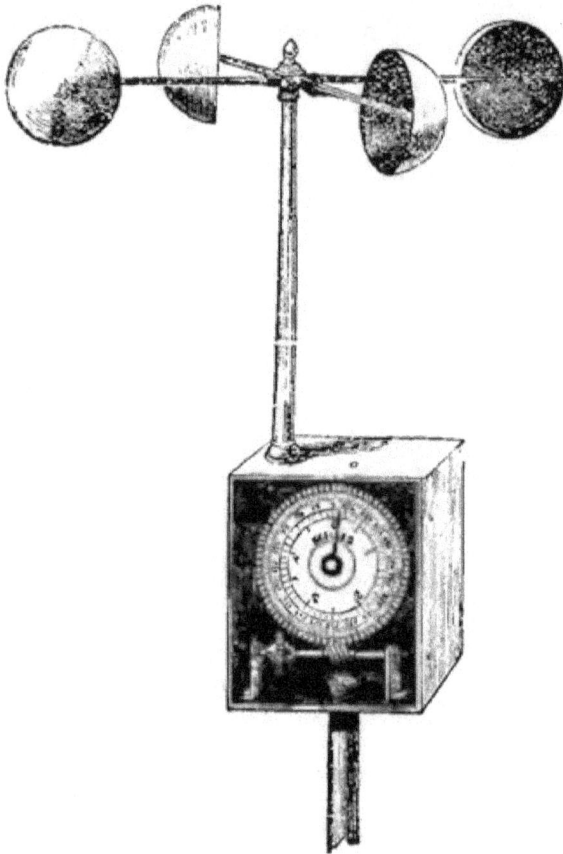

Improved Anemometer.   Scale about ¼.

It may be placed in any desired position by screwing the iron shaft supplied with it into the hole provided for the purpose, and fixing the apparatus on a pole or on an elevated stand, if possible, in an open space exposed to the *direct* action of the wind.

If, when placing the instrument, the hands stand at

0, the next reading will, of course, show the number of miles the wind has traversed; but, should they stand otherwise, the reading may be noted and deducted from the second reading, thus : Suppose the fixed index points to 2·5 and the movable index to 125, the reading after 12 hours may be 200 on the outer circle and 3·0 on the inner circle : these added together yield 203. By deducting the previous reading 127·5, we have the true reading—viz., 75·5 miles as the distance travelled.

Having obtained the velocity of the wind in this manner in miles per hour, the table on page 83 will enable the observer to calculate the pressure in pounds per square foot.

In this connection it should be noted that one of the results of the destruction of the Tay Bridge in 1879 was the appointment of a Commission to inquire into Wind Pressure.

WEATHER NOTATION.

The following letters are used to denote the state of the weather :

b denotes blue sky, whether with clear or slightly hazy atmosphere.
c   ,,   cloudy, that is detached opening clouds.
d   ,,   drizzling rain.
f   ,,   fog.
h   ,,   hail.
l   ,,   lightning.
m   ,,   misty, or hazy so as to interrupt the view.
o   ,,   overcast, gloomy, dull.
p   ,,   passing showers.
q   ,,   squally.
r   ,,   rain.
s   ,,   snow.
t   ,,   thunder.
u   ,,   ugly, threatening appearance of sky.
v   ,,   unusual visibility of distant objects.
w   ,,   wet, that is dew.

A letter repeated denotes much, as *rr*, heavy rain ; *ff*, dense fog ; and a figure attached denotes duration in hours, as 14*r*, 14 hours' rain.

By the combination of these letters all the ordinary phenomena of the weather may be recorded with certainty and brevity.

Examples.—*bc*, blue sky with less proportion of cloud ; *cb*, more cloudy than clear ; 2*rrllt*, heavy rain for two hours, with much lightning, and some thunder.

VELOCITY AND PRESSURE OF THE WIND.

The Pressure varies as the Square of the Velocity, or $P \propto V_2$. The Square of the Velocity in Miles per Hour multiplied by ·500 gives the Pressure in lbs. per square Foot, or $V_2 \times ·005 = P$. The Square Root of 200 times the Pressure equals the Velocity, or $\sqrt{200 \times P} = V$.

The subjoined Table is calculated from this data, by COL. SIR HENRY JAMES, of the Ordnance Survey Office.

Pressure in lbs. per Square Foot	Velocity in Miles per Hour	Pressure in lbs. per Square Foot	Velocity in Miles per Hour	Pressure in lbs. per Square Foot	Velocity in Miles per Hour	Pressure in lbs. per Square Foot	Velocity in Miles per Hour	Pressure in lbs. per Square Foot	Velocity in Miles per Hour
oz.		lbs.		lbs.		lbs.		lbs.	
0·08	1 000	6·75	36·742	17·75	59·581	28·75	75·828	39·75	89·162
0·25	1·767	7·00	37·415	18·00	60·000	29·00	76·157	40·00	89·442
0·50	2·500	7·25	38·078	18·25	60·415	29·25	76·485	40·25	89·721
0·75	3·061	7·50	38·729	18·50	60·827	29·50	76·811	40·50	90·000
1·00	3·535	7·75	39·370	18·75	61·237	29·75	77·136	40·75	90·277
2·00	5·000	8·00	40·000	19·00	61·644	30·00	77·459	41·00	90·553
3·00	6·123	8·25	40·620	19·25	62·048	30·25	77·781	41·25	90·829
4·00	7·071	8·50	41·231	19·50	62·449	30·50	78·102	41·50	91·104
5·00	7·905	8·75	41·833	19·75	62·849	30·75	78·421	41·75	91·378
6·00	8·660	9·00	42·426	20·00	63·245	31·00	78·740	42·00	91·651
7·00	9·354	9·25	43·011	20·25	63·639	31·25	79·056	42·25	91·923
8·00	10·000	9·50	43·589	20·50	64·031	31·50	79·372	42·50	92·195
9·00	10·606	9·75	44·158	20·75	64·420	31·75	79·686	42·75	92·466
10·00	11·180	10·00	44·721	21·00	64·807	32·00	80·000	43·00	92·738
11·00	11·726	10·25	45·276	21·25	65·192	32·25	80·311	43·25	93·005
12·00	12·247	10·50	45·825	21·50	65·574	32·50	80·622	43·50	93·273
13·00	12·747	10·75	46·368	21·75	65·954	32·75	80·932	43·75	93·541
14·00	13·228	11·00	46·904	22·00	66·332	33·00	81·240	44·00	93·808
15·00	13·693	11·25	47·434	22·25	66·708	33·25	81·547	44·25	94·074
		11·50	47·958	22·50	67·082	33·50	81·853	44·50	94·339
lbs.		11·75	48·476	22·75	67·453	33·75	82·158	44·75	94·604
1·00	14·142	12·00	48·989	23·00	67·823	34·00	82·462	45·00	94·868
1·25	15·811	12·25	49·497	23·25	68·190	34·25	82·764	45·25	95·131
1·50	17·320	12·50	50·000	23·50	68·556	34·50	83·066	45·50	95·393
1·75	18·708	12·75	50·497	23·75	68·920	34·75	83·366	45·75	95·655
2·00	20·000	13·00	50·990	24·00	69·282	35·00	83·666	46·00	95·916
2·25	21·213	13·25	51·478	24·25	69·641	35·25	83·964	46·25	96·176
2·50	22·360	13·50	51·961	24·50	70·000	35·50	84·261	46·50	96·436
2·75	23·452	13·75	52·440	24·75	70·356	35·75	84·567	46·75	96·695
3·00	24·494	14·00	52·915	25·00	70·710	36·00	84·852	47·00	96·953
3·25	25·495	14·25	53·385	25·25	71·063	36·25	85·146	47·25	97·211
3·50	26·457	14·50	53·851	25·50	71·414	36·50	85·440	47·50	97·467
3·75	27·386	14·75	54·313	25·75	71·763	36·75	85·732	47·75	97·724
4·00	28·284	15·00	54·772	26·00	72·111	37·00	86·023	48·00	97·979
4·25	29·154	15·25	55·226	26·25	72·456	37·25	86·313	48·25	98·234
4·50	30·000	15·50	55·677	26·50	72·801	37·50	86·602	48·50	98·488
4·75	30·822	15·75	56·124	26·75	73·143	37·75	86·890	48·75	98·742
5·00	31·622	16·00	56·568	27·00	73·484	38·00	87·177	49·00	98·994
5·25	32·403	16·25	57·008	27·25	73·824	38·25	87·464	49·25	99·247
5·50	33·166	16·50	57·445	27·50	74·181	38·50	87·749	49·50	99·498
5·75	33·911	16·75	57·879	27·75	74·498	38·75	88·034	49·75	99·749
6·00	34·641	17·00	58·309	28·00	74·833	39·00	88·317	50·00	100·000
6·25	35·355	17·25	58·736	28·25	75·166	39·25	88·600		
6·50	36·055	17·50	59·160	28·50	75·498	39·50	88·861		

This is the only table hitherto much in use for converting velocity into pressure, and was prepared by Smeaton and others. It does not, however, express the true relation, which has yet to be determined.

The Anemograph, or Self-Recording Wind Gauge, has for its object the registration of the velocity and direc-

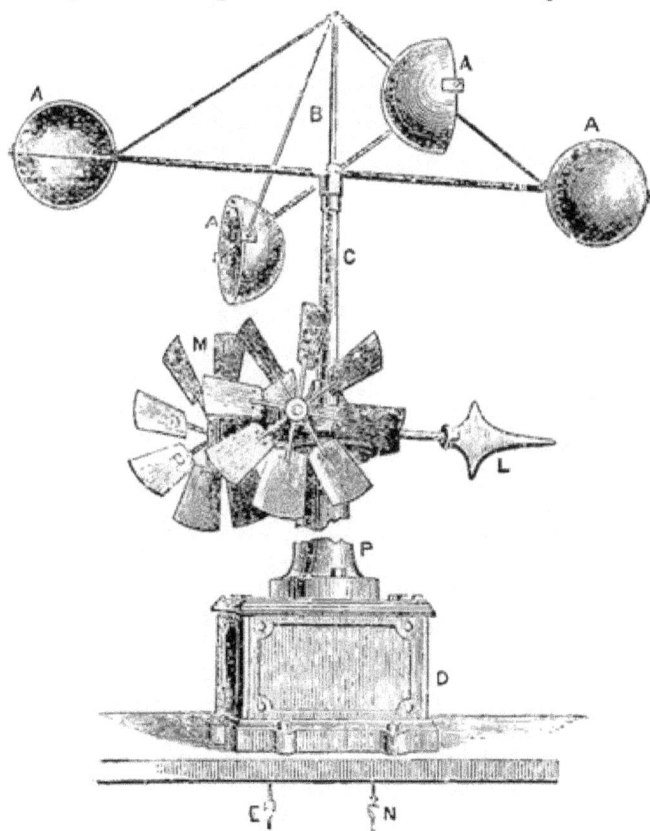

59.

Anemograph. Scale about $\frac{1}{15}$.
Portion for exterior of observatory.

tion of the wind from day to day. Figs. 59 and 60 show the form designed and arranged by Mr. Beckley, of the Kew Observatory, which has been adopted by the Meteorological Office.

It consists of a set of hemispherical cups and vanes, which are exposed on the roof of the house, and of the recording apparatus, which is placed inside the house.

The motion imparted to the hemispherical cups by the wind is communicated to the steel shaft B, which, passing through the hollow shaft C, and having at its lower end an endless screw, works into a series of wheels in the iron box D, which reduces the angular velocity 7,000 times. At the required distance the motion, having emerged at E, is connected with F, where, by means of bevelled wheels, it moves the spiral brass registering pencil C, which is arranged so that each revolution records 50 miles of velocity on the prepared paper H.

The direction of the wind is indicated by the arrow L,

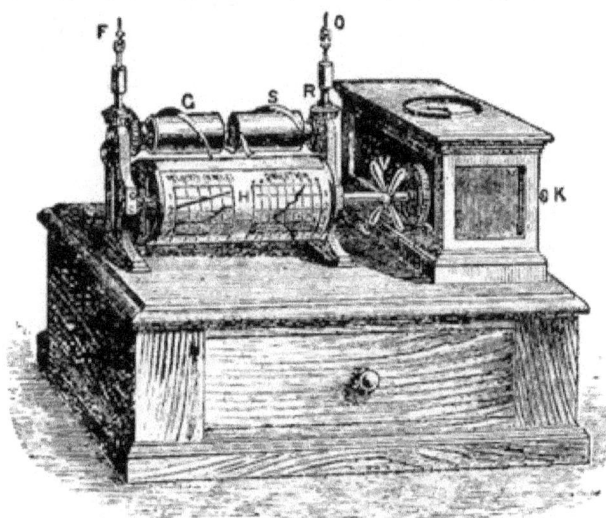

60.

Anemograph. Scale about $\frac{1}{20}$.
Portion for interior of observatory.

which is kept in position by the fans M. These communicate, by an endless screw and train of wheels, through the shaft C and the box D to the recording apparatus, consisting of a spiral brass pencil, which in one revolution records variations through the cardinal points of the compass, on the same prepared paper as that which receives the record of velocity.

The paper is held on the drum by two small clips, and may be readily changed, by unclamping the cross V, without disturbing the drum or any other part of the instrument.

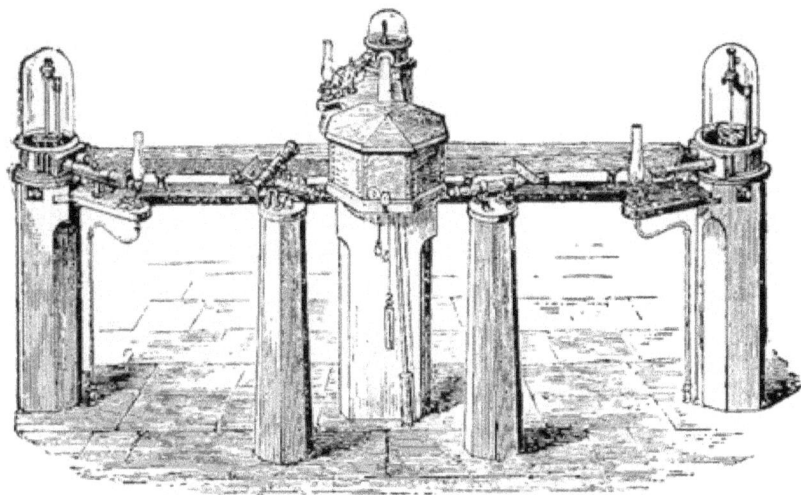

61.

Self-recording Magnetometer, Kew Observatory.

VI.—ELECTRIFICATION.

William Gilbert, a physician of Colchester, first showed in 1600 that the earth as a whole has the properties of a magnet, and consequently that the directive action exerted by it upon a compass needle represents only a special case of the mutual action of two magnets. In 1845, Faraday established the fact that susceptibility to magnetic force is not, as was generally believed, confined to iron, nickel, and a few other substances, but is a property of all substances. According to Balfour Stewart, auroræ and earth currents may be regarded as secondary currents resulting from changes in the earth's magnetism. Magnetic phenomena are included under the general term terrestrial magnetic elements, and consist of magnetic declination, inclination, and intensity.

These are for convenience determined separately; the first by an instrument called a *Declinometer*, and the second by an *Inclinometer* or *Dipping Needle*. The Declinometer is also made to serve the additional purpose of measuring the *intensity* of the earth's magnetic force, which it effects on a principle similar to that by which the force of gravity is determined by the oscillations of a pendulum of known length on any given portion of the earth's surface. The declinometer needle is made to oscillate, and the number of oscillations in a given time counted; due allowance being made for the strength of the needle, it is obvious that the force which restores the needle to rest can be estimated. To ascertain the angle of *declination*, the zero line of the compass card is made to coincide with the geographical north and south line; and the angle which the direction of the needle makes with this line is then read off on a graduated circle over which the needle turns. The magnetic *inclination* or *dip of the needle* is estimated by observing the inclination to a horizontal plane of a needle turning on the vertical plane which passes through the magnetic north and south points.

Fig. 62 shows a simple form of magnetic needle suspended on a fine steel point, which is supported by a brass stand; the addition of a graduated circle would constitute such an arrangement a Declinometer.

Fig. 63 gives the appearance of the dipping needle, or Inclinometer, and Fig. 65 an arrangement by which both kinds of terrestrial as well as local attraction may be shown.

63.

64.

These components of the earth's magnetism undergo not only an annual but a daily and even hourly variation, apparently connected in some occult manner with the frequency of the sun's spots. The needle sometimes suffers such exceptional perturbations as to suggest the

idea of a magnetic storm. These disturbances are usually
accompanied (in polar regions) by luminous phenomena
called auroræ. Continuous automatic record of them,
therefore, is of great value, as facilitating inductive research which may lead to valuable practical results.

Accordingly the Royal Society have adopted for the
Kew and other observatories the form of Magnetograph,
or Self-recording Magnetometer, shown at Fig. 61, by
means of which the variations just referred to are registered by the oscillations of three magnets on photo-
graphically prepared paper, stretched on a drum revolved
by clockwork.

One magnet is suspended in the magnetic meridian
by a silk thread, and, by the aid of a mirror attached,
it describes on the cylinder, moved by clockwork in the
centre pier, all the variations in the magnetic *declination*.

The other two components of the magnetic force of the
earth are given by the other magnets. That recording
the vertical variations rests on two agate edges under a
glass shade, while the horizontal component magnet is
suspended by a double silk thread, under the shade to
the right of the picture, being retained by the tension
of the thread in a position nearly at right angles to the
magnetic meridian.

The clock box in the centre covers the three revolving
cylinders bearing the sensitive photographic paper, and
to each magnet is attached a semicircular mirror, which
reflects the rays from a gas jet to one of the cylinders,
and thus describes by a curved line the oscillations of
the magnet. A second semicircular mirror is *fixed* to
the pier on which the instrument stands, and consequently describes a straight line, or zero, from whence
the curves are measured.

To avoid errors attending sudden changes of temperature, underground vaults are always chosen for magnetic
observations, and also on account of light being more
easily and perfectly excluded.

ATMOSPHERIC ELECTRICITY.

Since the performance of Franklin's famous kite experiment, by which he determined the identity of lightning with the electrical discharge from a machine, much attention has been devoted, not only to that form of atmospheric electricity which displays itself in the thunder-cloud, but to the electric condition of the air in all states of the weather. These researches have established the fact that the air is always in an electrical condition, even when the sky is clear and free from thunder-clouds. The instruments employed for ascertaining the kind and intensity of atmospheric electricity are called Electroscopes. Fig. 65 shows a modification of Saussure's Electroscope, the basis of which is a narrow-mouthed flint glass bottle with a divided scale to indicate the degree of divergence of the gold leaves or straws. To protect the lower part from rain, it is covered by a metallic shield about five inches in diameter. Bohnenberger's Electroscope indicates the presence and quality of *feeble* electric currents. Peltier's Electrometer yields the same result by the deflection of a magnetic needle. This latter has been in use at Brussels for thirty years, and at Utrecht for twenty years, and is highly recommended.

Singer's Atmospheric Electroscope is an efficient form of the instrument in which an ordinary gold-leaf electrometer has attached to its circular brass plate a brass rod two feet in length, with a clip at its upper ex-

65.

Electroscope.
Scale about ¼.

tremity to receive a lighted paper or cigar fusee. The electricity of the air in immediate contact with the flame, causes, by induction, electricity of the opposite nature to accumulate at the upper extremity, where it is constantly carried off by the convection currents in the flame, leaving the conductor charged with the same kind and power of electricity as that contained in the air at the time of the experiment. The principle of this method was initiated by Volta, and has been extended and applied by Sir William Thomson in his Water-dropping Collector, which consists of an insulated cistern from which water escapes through a jet so fine that it breaks into drops immediately after leaving the nozzle of the tube. The result of this is that in half a minute from the starting of the stream the can is found to be electrified to the same extent as the air at the point of the tube. The scale value of each instrument has to be separately determined by repeated comparative experiments, and involves much delicacy of manipulation.

It is chiefly important for the ordinary observer to know that the occurrence of thunder and lightning should be always noted in the column headed " Remarks."

The destructive effects of lightning are too well known to need description here; the means, however, by which these may be averted demand a brief notice. Lightning when discharged from a cloud will always choose the better of any two conductors which may present themselves. The *stone* of a church steeple and the *wood* of a ship's mast are bad conductors, but a galvanized iron wire rope is the best possible conductor, and accordingly this material is now generally employed for the purpose. A lightning conductor consists of three parts: 1, the

66.
Lightning
Conductor.
Scale about $\frac{1}{15}$.

rod, which extends beyond the summit of the building, 2, the conductor, which connects the rod with the underground portion, and 3, the part underground. The connection between each of these must be absolutely perfect, or the conductor will be faulty. The top is usually of solid copper tipped with platinum (Fig. 66), the body of galvanized iron rope, so as to adapt itself to the inequalities of the building and yet have no sharp turns in it, while the part underground is of solid iron rod. This latter portion should extend straight underground for two feet, and being bent at right angles away from the wall, should rest in a horizontal drain 10 to 15 feet long filled with charcoal, and be again bent downwards into a well of water. Should water not be available, it should rest in the centre of a hole 15 feet deep and 10 inches in diameter, tightly packed with charcoal, which, while conducting the electricity from the rod into the earth, serves also to preserve the iron from rusting.

OZONE.

The atmosphere, besides holding the vapour of water diffused throughout its mass, contains also minute traces of carbonic acid and ammonia, and a very remarkable substance called Ozone. Oxygen, one of the component gases of the atmosphere, is capable of existing in two conditions; one in which it is comparatively passive, and another in which it possesses exceptional chemical activity, dependent apparently upon its electrical condition, and in which state it possesses a peculiar smell which has caused it to be named ozone.* The characteristic odour is always observable near a powerful electric machine when it is being worked, near a battery used for the decomposition of water, and in the air after the passage of a flash of lightning. Its presence is most marked near the sea-coast, and in localities remarkable for their salubrity; and on account of its influence on health, it has been proposed by Schonbein and others

* Greek *ozo*, I smell.

to include ozonometrical observations with the ordinary meteorological observations.

Although in minute quantities it is favourable to health, when existing in undue proportion it irritates the mucous membrane of the nose and throat, producing painful sores. It attacks india-rubber, bleaches indigo, and oxidizes silver and mercury, differing in all these points from ordinary atmospheric oxygen.

The chemical energy it possesses (which exceeds that of ordinary oxygen as much as the latter exceeds atmospheric air as an oxidizing agent) affords the means of ascertaining its presence and quantity. It liberates iodine from its combination with potassium, and free iodine colours starch a deep blue.

Schonbein, the discoverer of ozone, found that when strips of paper previously saturated with starch and iodide of potassium and dried were exposed freely to the air but protected from rain and the direct action of the sun, they underwent a peculiar discoloration (when immersed in water) after an exposure of 24 hours. A scale of tints numbered from one to ten afforded the means of comparative observation, and thus the Ozonometer was constructed, and a means established of registering the amount of ozone in the air of various localities from day to day.

67.

Ozone Cage.

Scale about ¼.

Schonbein also observed that the proportion of ozone was largely augmented after heavy falls of snow. For the exposure of the ozone papers, an ozone cage is employed, as shown at Fig. 67.

Ozone may be prepared artificially as a disinfectant by cautiously mixing without friction or concussion equal parts of peroxide of manganese, permanganate of potash, and oxalic acid. For a room containing 1,000 cubic feet, two teaspoonfuls of the powder, placed in a dish and moistened with water occasionally, will develop the ozone and disinfect the surrounding air without producing cough.

The most important and interesting series of facts, however, connected with ozone are those established by the researches of M. Houzeau, who states :—

1. That country air contains an odorous oxidizing substance, with the power of bleaching blue litmus, without previously reddening it, of destroying bad smells, and of bluing iodized red litmus.

2. That this substance is ozone.

3. That the amount of ozone in the air at different times and places is variable, but this is at most $\frac{1}{700,000}$ of its volume, or 1 volume of ozone in 700,000 of air.

4. That ozone is found much more frequently in the country than in towns.

5. That ozone is in greatest quantity in spring, less in summer, diminishes in autumn, and is least in winter.

6. It is most frequently detected on rainy days, and during great atmospheric disturbances.

7. That atmospheric electricity is apparently the great generator of ozone.

The subject is one of great interest in its bearings on health, and opens a wide field of scientific research, as may be inferred from the opinion expressed by the Vienna Congress, which is that " the existing methods of determining the amount of ozone in the atmosphere are insufficient, and the Congress therefore recommends investigations for the discovery of better methods."

Mr. Lowe has published the valuable weather warnings tabulated on page 94, which are interesting as showing from a given number of observations the value of each phenomenon :—

		No. of observations.	Followed in 24 hours by	
			F	Rain.
DEW.	Dew profuse	241	196	43
	Dew from 1st April to 30th Sept.	185	161	24
	Dew from 1st Oct. to 30th March	56	37	19
CLOUDS.	White stratus in the valley ...	229	201	28
	Coloured clouds at sunset ...	35	26	9
SUN.	Solar halos	204	133	71
	Sun red and shorn of rays ...	34	31	3
	Mock suns	35	19	6
	Sun shone through thin cirro-stratus...	13	6	7
	Sun pale and sparkling... ...	51	27	24
FROST.	White frost	73	59	14
MOON.	Lunar halos	102	51	51
	Mock moons	9	7	2
	Lunar burr	64	47	17
	Moon shining dimly	18	12	6
	Moon rose of a red colour ...	8	7	1
STARS.	Falling stars abundant	85	65	20
	Stars bright	83	64	19
	Stars dim	54	32	22
	Stars scintillated	14	12	2
AURORA.	Aurora borealis	76	49	27
ANIMALS.	Bats flying about in the evening	61	45	16
	Toads in the evening	17	12	5
	Landrails clamorous	14	13	1
	Ducks and geese noisy	10	7	3
	Spiders hanging on webs in the evening	8	5	3
	Fish rise in the lake	15	9	6
SMOKE.	Smoke rising perpendicularly ...	6	5	1

Among the animals whose movements give weather warnings few are more trustworthy than the leech. The reader may verify this by placing one in a broad glass bottle, tied over with perforated leather, or bladder. If placed in a northern aspect, the leech will be found to behave in the following manner :—

1. On the approach of fine or frosty weather, according to the season, it will be found curled up at the bottom. 2. On the approach of rain, snow, or wind, it will rise excitedly to the surface. 3. Thunder will cause it to be much agitated, and to leave the water entirely.

PERIODS.—M. Köppen states, as the result of his examination into the chances of a change of weather, that *the weather has a decided tendency to preserve its character*. Thus, at Brussels, if it has rained for nine or ten days successively, the *next* day will be wet also in four cases out of five ; and the chance of a change decreases with the length of time for which the weather *from* which the change is to take place has lasted.

In the case of temperature for five-day periods, the same principle holds good ;* for if a cold five-day period sets in after warm weather, we can bet two to one that the next such period will be cold too ; but if the cold has lasted for two months, we can bet nearly eight to one that the first five days of the next month will be cold too. The chance of change is, however, greater for the five-day periods than for single days. Similar results follow for the months, but here again the chance of change shows an increase.

"If we revert to the instance first cited, that of rain, the result is, *not* that if it once begins to rain the chances are in favour of its never ceasing ; all that is implied is, that the chances are against its ceasing on a definite day, and that they increase with the length of time the rain has lasted. The problem is similar to that of human life : the chance of a baby one year old living another year is less than that of a man of thirty.

"The practical meaning of all this is, that although we know that a compensating anomaly for all extraordinary weather exists somewhere on the earth's surface, *e. g.*, the very common case of intense cold in America, while we have a mild winter in Britain, there is no season as yet ascertained to anticipate that this compensation will occur at any given place during the year. In other words, when definite conditions of weather have thoroughly established themselves, it is only with great difficulty that the courses of the atmospheric currents are changed."

* "Recent Progress in Weather Knowledge," by R. H. Scott, F.R.S.

To bring within the limits of a popular pamphlet a notice of the various phenomena classed under the head of Meteorology, it has been necessary to exercise the utmost brevity. Brief, however, as the treatment has been, reference has been made to the sciences of Heat, Light, Electricity, Magnetism, Gravitation, Astronomy, Chemistry, Geography, and Geology, thus corroborating the testimony of Sir John Herschel, who states that "it can hardly be impressed forcibly enough on the attention of the student of nature that there is scarcely any natural phenomenon which can be fully and completely explained in all its circumstances without a union of several— perhaps of all—the sciences ; and it cannot be doubted that whatever walk of science he may determine to pursue, impossible as it is for a finite capacity to explore all with any chance of success, he will find it illuminated in proportion to the light which he is enabled to throw upon it from surrounding regions. But, independently of this advantage, the glimpse which may thus be obtained of the harmony of Creation, of the unity of its plan, of the theory of the material universe, is one of the most exalted objects of contemplation which can be presented to the faculties of a rational being. In such a general survey he perceives that science is a whole whose source is lost in infinity, and which nothing but the imperfection of our nature obliges us to divide. He feels his nothingness in his attempts to grasp it, and he bows with humility and adoration before that Supreme Intelligence who alone can comprehend it, and who 'in the beginning saw everything that He had made, and behold it was very good.'"

WILLIAM RIDER AND SON, PRINTERS, LONDON